ENERGY PRICES AND PUBLIC POLICY

A Statement by the Research and
Policy Committee of the Committee
for Economic Development
and
The Conservation Foundation

July 1982

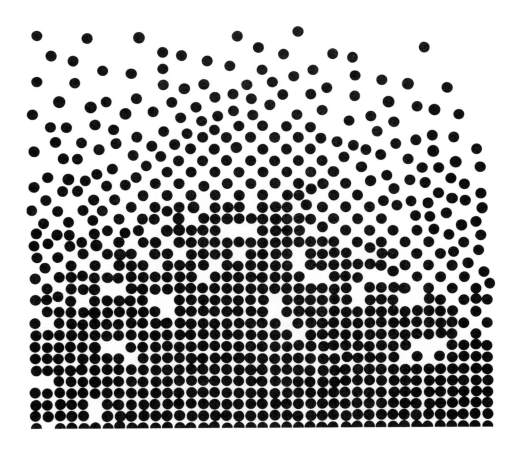

Library of Congress Cataloging in Publication Data

Committee for Economic Development. Research and Policy
  Committee.
  Energy prices and public policy.

  1. Energy policy — United States. 2. Fuel — United
States — Price policy. 3. Energy policy. 4. Fuel —
Price policy. I. Conservation Foundation. II. Title.
HD9502.U52C65   1982   333.79'0973   82-7428
ISBN 0-87186-075-9 (pbk.)   AACR2
ISBN 0-87186-775-3 (lib. bdg.)

First printing: July 1982
Paperbound: $7.50
Library Binding: $9.50
Printed in the United States of America by The Kearny Press, Inc.
Design: Stead, Young & Rowe, Inc.

COMMITTEE FOR ECONOMIC DEVELOPMENT
477 Madison Avenue, New York, N.Y. 10022
1700 K Street, N.W., Washington, D.C. 20006

CONTENTS

ENERGY PRICES AND PUBLIC POLICY

RESPONSIBILITY FOR CED STATEMENTS ON NATIONAL POLICY

The Committee for Economic Development is an independent research and educational organization of two hundred business executives and educators. CED is nonprofit, nonpartisan, and nonpolitical. Its purpose is to propose policies that will help to bring about steady economic growth at high employment and reasonably stable prices, increase productivity and living standards, provide greater and more equal opportunity for every citizen, and improve the quality of life for all. A more complete description of CED is to be found on page 88.

All CED policy recommendations must have the approval of the Research and Policy Committee, trustees whose names are listed on page vii. This Committee is directed under the bylaws to "initiate studies into the principles of business policy and of public policy which will foster the full contribution by industry and commerce to the attainment and maintenance" of the objectives stated above. The bylaws emphasize that "all research is to be thoroughly objective in character, and the approach in each instance is to be from the standpoint of the general welfare and not from that of any special political or economic group." The Committee is aided by a Research Advisory Board of leading social scientists and by a small permanent professional staff.

The Research and Policy Committee is not attempting to pass judgment on any pending specific legislative proposals; its purpose is to urge careful consideration of the objectives set forth in this statement and of the best means of accomplishing those objectives.

Each statement is preceded by extensive discussions, meetings, and exchanges of memoranda. The research is undertaken by a subcommittee, assisted by advisors chosen for their competence in the field under study. The members and advisors of the subcommittee that prepared this statement are listed on page viii.

The full Research and Policy Committee participates in the drafting of findings and recommendations. Likewise, the trustees on the drafting subcommittee vote to approve or disapprove a policy statement, and they share with the Research and Policy Committee the privilege of submitting individual comments for publication, as noted on pages vii and 82 of this statement.

Except for the members of the Research and Policy Committee and the responsible subcommittee, the recommendations presented herein are not necessarily endorsed by other trustees or by the advisors, contributors, staff members, or others associated with CED.

CED RESEARCH AND POLICY COMMITTEE

PURPOSE OF THIS STATEMENT

This report is the product of an unusual collaboration between interests commonly considered adversary, namely business leaders and conservation groups. It represents an agreement about the national interest and the role of energy policy. At a time when energy prices are falling and the "energy crisis" is said to have passed, we wish to focus the attention of public officials and opinion leaders on an urgent imperative: the need to recognize that market pricing should be the chief mechanism for achieving energy conservation and allocating and stimulating energy supplies.

The many reasons for our policy statement are set out in detail in this report. Fundamentally, we believe that the nation must once and for all adapt its laws, its economy, its industrial investment and its product design to the realities of world energy supply and demand. Our failure to do so earlier in the 1970s, while many of our competitors in other countries were adapting and retooling their industrial economies to world energy prices, has cost this nation dearly. Although the United States has been moving in the right direction for the past few years, it has done so in fits and starts, without deeply accepted public conviction, and the result has been only a partial lifting of controls on energy prices. With this report, The Conservation Foundation and the Committee for Economic Development (CED) hope to make the case to both the business and environmental group leaderships that the nation should continue to move toward market-determined energy prices and that it should not look back.

A COLLABORATION BETWEEN BUSINESS LEADERS AND CONSERVATIONISTS

Energy policy is a highly polarizing subject on which feelings run high and differences between interest groups are pronounced. This policy statement, *Energy Prices and Public Policy*, is tangible evidence that significant agreement is possible between business executives and conservationists. It is the product of an unusual partnership between the Committee for Economic Development, an organization of business leaders, and The Conservation Foundation, an organization of conservationists.

While this project is unusual, it is not unique. Rather, it is part of a larger movement throughout the country involving the resolution of conflict between business and environmental organizations. The Conservation Foundation itself has organized several "Dialogue Groups" on subjects ranging from toxic-substance control to forest management and low-level radioactive-waste disposal. This joint policy statement adds strength to that movement. It forcefully demonstrates that business executives and conservationists can meet in a nonadversarial forum, reconcile at least some of their differences, and find important areas of agreement, without resorting to prolonged disputes in the courts, the Congress, and the state legislatures. Quite the opposite of selling out or giving in to one another, policy discussions leading to agreements serve to generate more intelligent and analytical arguments even in those areas where the parties are unable to reach agreement. Ultimately, better public-policy decisions can emerge, even where positions differ.

In the field of energy policy, both business leaders and environmentalists have found their own positions developing over time. Within CED, an important milestone was the 1979 publication of *Thinking Through the Energy Problem* by Thomas C. Schelling as a supplementary paper. In that volume, the conceptual framework for the role of market prices in the energy field was put forth, providing an opportunity for leaders of the business community to test their own beliefs and policies against the thinking of a leading academic economist.

In a similar fashion, environmental groups have moved from a nervous unease concerning the role of economic theory in energy policy towards an understanding of how proper pricing can actually substantially benefit the environment. The resource-conservation effects of proper pricing of scarce materials—particularly if side effects are properly quantified and included in the market price—can achieve many of the environmental and conservation goals that other, clumsier mechanisms often fail to provide.

For each side in this policy discussion, reaching agreement meant a reexamination of traditional assumptions and alliances. Many companies are large consumers of energy; a few earn the bulk of their revenues from providing it. Within the business community there are clearly winners and losers from any change from more traditional policies. Similarly, within the environmental community, discussion of market prices strains the long-standing cordial relationship with consumer groups.

The series of extensive discussions—with their tentative positions, drafts, restatements, and then, finally, acceptance or withdrawal—that led up to the production of this joint policy statement gradually made clear to both sides the mutual benefits of a more market-oriented energy-pricing policy in this country.

TOWARD MARKET SOLUTIONS OF ENERGY PROBLEMS

This statement appears at a propitious time. The momentum toward a market system for energy pricing is not as strong as many would believe.

Natural gas prices remain under control; electricity prices have yet to move significantly toward market-determined pricing; and planning that is necessary to preserve the benefits of the market-pricing system in the event of another oil shortfall emergency is not being done. More poignantly, even the low-income consumers have not been addressed, nor are the hardships that additional price deregulation might cause the poor being faced. In short, there remains work to be done, and this statement proposes in concrete terms some answers to the decisions that lie ahead.

RESPONSIBILITIES AND ACKNOWLEDGEMENTS

In accordance with the procedures of the Committee for Economic Development, outlined on page vi, a draft of this statement was submitted for approval to the CED Research and Policy Committee; CED trustee members are entitled to submit memoranda of comment, reservation, or dissent, which are published in the back of this document. Procedures of The Conservation Foundation do not require individual trustees to take responsibility for its policy statements. The Conservation Foundation Board of Trustees has approved the report in principle. (A listing of members of CED's Research and Policy Committee appears on page vii. The Conservation Foundation's Board is listed in the back of this book.)

Special commendation is due to Thomas C. Schelling, Lucius N. Littauer Professor of Political Economy, Harvard University, and Grant P. Thompson, Senior Associate at The Conservation Foundation, the talented project co-directors for CED and The Conservation Foundation, respectively, for the teamwork they displayed in guiding this project through successive drafts to produce the final report. We also extend our gratitude to the members of the joint subcommittee from both organizations (listed on page viii) who met in a spirit of cooperation, trust, and mutual respect, determined to find areas of agreement.

Finally, we are indebted to the Ford Foundation and the John M. Olin Foundation for their generous support of this project.

Henry B. Schacht, Co-Chairman
Chairman
Cummins Engine Company, Inc.

William K. Reilly, Co-Chairman
President
The Conservation Foundation

Franklin A. Lindsay
Chairman, CED Research and
  Policy Committee
Chairman, Executive Committee
Itek Corporation

John A. Bross
Chairman, Executive Committee
The Conservation Foundation

EXECUTIVE SUMMARY

POLICY TOWARD ENERGY PRICES (CHAPTER 1)

Energy prices, together with all other prices, have two main jobs to do in our kind of economic system. One is to allocate energy resources to their most valued uses. The second is to serve as sources of income and arbiters of what we get for the money we spend. But energy pricing in the United States has also often been used or proposed as an instrument for goals as diverse as assisting the poor, controlling inflation, protecting the environment, changing residential patterns, and reducing highway fatalities.

It is increasingly clear, however, that controlling energy prices has neither increased energy supplies nor made significant contributions to reducing poverty, curbing inflation, or achieving the wide variety of economic and social goals that controlled energy prices have been supposed to aid.

In fact, controls on the price of energy have fueled excessive consumption, inhibited exploration and investment, and undervalued technologies that would help ease the energy supply problem and promote conservation. In addition, artificially low energy prices have not allowed industry and consumers to make decisions based on the actual cost of energy and have locked this country into an even-higher-cost energy future.

We favor increased reliance on the market system in pricing energy so that production, consumption, and conservation decisions can reflect real costs.

This policy statement examines the principles underlying the price system, both those that make competitive prices attractive mechanisms for energy policy and those that justify special interventions. It closely examines two of the most frequently cited arguments for controlling energy prices—protecting low-income consumers and fighting inflation—and concludes that energy policy is a poor means of achieving either goal. The report recommends ways in which electricity and natural gas pricing can be more market-oriented and presents a market strategy for dealing with oil emergencies.

ENERGY-PRICING POLICY AND LOW-INCOME CONSUMERS (CHAPTER 2)

A more market-oriented system of pricing energy will mean that energy prices are likely to rise, perhaps even sharply. Although all Americans will feel the impact of higher prices, no one group should suffer unduly. We believe that any national commitment to market pricing of energy must be accompanied by an equally strong commitment to seeing that low-income consumers, who spend proportionately more of their income on energy, do not suffer more than other groups.

Any policy of market pricing of energy must take the impact on the poor into account and must directly address the welfare needs that market pricing will create. However, we believe that it is critically important to separate these two issues; energy policy should not be used as a substitute for an adequate welfare policy for taking care of those in genuine need. We believe that the poor ought to receive adequate welfare increases to help compensate them for what we calculate as a $3-to-$7-billion decrease in their annual real income caused by energy price increases over the past eight years.

Although additional welfare benefits may be costly, they are a very small price to pay to break the political deadlock over decontrol. And adequate benefits represent a measure of decency and fairness to the poor to cushion the impact of the burden they disproportionately bear as the country moves to a more rational energy policy.

We also recommend increasing efforts to upgrade the energy efficiency of the homes of low-income consumers. By investing relatively small amounts of money in improving the homes and energy-consuming appliances of the poor, government can both decrease the need for energy-related welfare assistance and help reduce overall national energy consumption. Existing weatherization programs should be expanded to allow structural repairs to dwellings and allow for repair and replacement of such capital items as furnaces, water heaters, and refrigerators.

ENERGY PRICES AND INFLATION (CHAPTER 3)

Rising fuel prices have been a major element in U.S. inflation in the past eight years. The cumulative effect of oil price shocks has enhanced the underlying rate of inflation and has thus generated a substantial inflationary effect beyond that caused by the increased oil expenditures themselves.

This chapter acknowledges that the recommendations for price decontrol could expose the economy to more inflationary pressures in the short run. But we also believe that the alternative of continued energy price controls would have even worse effects over time.

After examining the effectiveness of price controls in dealing with inflation, the chapter concludes that controls have not insulated the country from the rising cost of foreign oil and have indeed, by stimulating consumption, increased that cost. Controls have only postponed the inevitable increases in domestic oil and gas costs. We conclude that they have done more harm than good by encouraging consumption and discouraging production and conservation.

Although price regulations can appear to soften the effects of rising energy prices by phasing in increases, regulations that postpone inevitable increases weaken incentives to conserve and produce and, in the end, lead to even higher prices and higher inflation.

ELECTRICITY-PRICING POLICY (CHAPTER 4)

Public control of utility rates was established to allow natural monopolies to operate, subject to government oversight. But it has also had the effect of permitting regulated rate structures that have insulated both the public and the power companies from the real cost of producing electricity.

Most state public utility commissions set rates that are both averaged and based on historical, embedded cost, so that no customer is charged today's cost of producing the electricity that is being used at a given time. We recommend that national policy should have the goal of encouraging a shifting of electricity pricing in the direction of a pricing system that is based on the replacement costs of electricity for all customers. We recognize that there are a number of barriers, practical as well as political, to such pricing of electricity. The major barrier is that if companies were permitted to charge rates based on the cost of the most recently constructed plant, there would be an extremely large increase in utility revenues as consumers paid for electricity at the cost of producing it from the most expensive source, even though some of it came from older, cheaper plants. That barrier would be reduced by some means of taxing away any revenue in excess of expenses (including realistic depreciation) and an adequate rate of return, although designing and handling such a tax program would pose difficulties.

Complete replacement-cost pricing of electricity may not be feasible now, but nevertheless, there are a number of steps that the federal government could take to strenghten the ability of regulatory bodies to reform rates in the direction of more marketlike rates. Federal directives that require utilities to analyze their costs so that replacement costs are revealed have already served as a valuable basis for future reform. The federal government should now provide support to states willing to experiment with allowing more market-determined rates.

We also endorse, as a long-term goal, moving toward deregulating as much of the electricity utility industry as possible, thereby encouraging electricity generation from a variety of sources. We believe that some dispersal of generating capacity would be helpful, provided that it properly valued the reliability of our interconnected system. Such dispersal would enable the local utility to buy power from a sizable number of producers, give industry more flexibility, inject more competitive discipline, and encourage more economical operations.

The policy statement acknowledges that public utility commissions must contend with a variety of concerns in establishing rate bases and that they frequently must balance the competing concerns of earnings levels, fairness, inflation, price discrimination, and the appropriate reliability and quality of service. Public utility commissions are further hampered by rate-setting precedents that were established in times of long-term price stability. Under the existing regulatory system, most electric rates fall far short of replacement costs. Most electric rates today are inadequate to maintain the financial stability of important parts of the electric power industry. Hence, public utility commissions should directly address the question of adequate rates of return and cash flow for utilities.

Over time, the guiding principle ought to be to move electricity pricing in the direction of a system based on replacement costs for all customers.

NATURAL GAS PRICING POLICY (CHAPTER 5)

Faster deregulation of natural gas prices may result in increased production and certainly will reward greater conservation.

DEREGULATION AT THE WELLHEAD.   Current price regulations for natural gas under the Natural Gas Policy Act (NGPA) are both unrealistic and unwise.

- The price of *new conventional gas* (gas discovered after February 1977) is allowed to increase at the inflation rate plus a series of increases designed to reach the assumed world oil price of $15 a barrel (equivalent to $2.70 per million British thermal units) in 1985. The new gas will be completely decontrolled in 1985, with a possible sharp price jump from $2.70 per million Btus to whatever it can command in the open market.

- *Old gas* (i.e., gas in production at the beginning of 1977), on the other hand, is allowed to increase at its 1978 price plus adjustments for the rate of inflation. But the price of old gas will remain controlled beyond 1985, until its supplies are exhausted, possibly by the 1990s. The scheduled decontrol of natural gas under the NGPA is thus only a partial decontrol.

- *Unconventional gas* (from sand, shale, coal, deep geological formations, and so on) is currently not regulated, and no price regulations should be imposed.

We endorse decontrol covering all categories of natural gas. We have carefully weighed the argument for immediate decontrol of all categories of natural gas against that for phased partial decontrol in 1985 as prescribed under the NGPA. Under the current rate structure and pace of price increases in world oil, there is bound to be a sharp price hike in 1985, when new gas is decontrolled. We believe that delaying that increase makes little or no sense and that it is better to take a price shock sooner rather than later. We believe the economy will adjust to deregulation of natural gas prices, just as it adjusted to the deregulation of oil prices. Prompt decontrol of all types of natural gas will have a major national benefit in helping to order (or reorder) investment decisions with regard to new supply technologies or transmission facilities.

We believe that decontrol of natural gas is so important that if an excise tax on the additional revenues generated by decontrol were the price of achieving this goal, it would be a price worth paying. We do not favor such a tax, but if it becomes necessary to impose it, the tax should apply only to old gas. It should be phased out with the depletion of old gas itself, and the proceeds should go into general revenues instead of into dedicated trust funds.

If full and immediate decontrol is not acceptable, the government should at least set a short period of time during which natural gas prices would rise by substantial amounts each year until full decontrol is achieved. Because of the importance of moving prices quickly, it is vital that such a period be truly short.

NEW RATE STRUCTURES. Traditionally, state utility commissions have based their rate structures on average accounting costs. We urge that state commissions and distribution companies instead move toward rates that reflect the economic value of natural gas to users. We note that pipeline rate regulation may also require reform.

USES OF NATURAL GAS. Government regulations that limit the uses to which natural gas may be put are unnecessary and should be eliminated.

COPING WITH OIL EMERGENCIES (CHAPTER 6)

In a severe oil emergency, the overriding concern must be cooperation with other importing countries through the International Energy Agency (IEA), to which the United States is committed by treaty. Failure of the principal IEA countries to discipline themselves in sharing existing supplies would invite economic disaster and severely jeopardize the Atlantic alliance.

Domestically, prices should be the essential mechanism for allocating petroleum supplies. The emergency-reduced supply of imported oil should be allocated to domestic refiners by a government auction of import licenses, and the proceeds to the government from the auction should be promptly returned to the public.

We recommend that gasoline prices be allowed to rise to market levels, which in an emergency might be very high, and that taxes be levied to absorb a high percentage of the price increase. To avoid high income transfers and a major loss of purchasing power, the tax should be rebated to the public.

Gasoline rationing is a defensible alternative in an emergency if ration coupons are fully marketable. But a tax-rebate system can accomplish approximately what "white market" ration coupons could accomplish and can do so with less administrative burden and delay. Most effective would be a high tax on the crude oil price increases coupled with the auctioning of both import licenses and access to the strategic petroleum reserve. Such an auction would avoid distortions in the treatment of gasoline and other fuels. The existing windfall profits tax on crude oil could be the basis for an effective oil tax-rebate system.

Whichever of these two approaches is chosen, careful planning and preparatory measures in advance of any energy emergency are imperative to avoid the huge macroeconomic impact of greatly elevated prices or the enormity of controls and allocations.

The price system should not be abandoned during an emergency.

CHAPTER ONE
POLICY TOWARD
ENERGY PRICES

Why does there need to be a policy toward energy prices?

Prices are the signals that coordinate economic decisions. They are measures of worth, indexes of scarcity, incentives to economy, criteria for substitution, and common denominators for transactions. They mediate the billions of economic decisions that get made every day. In infinite detail and comprehensiveness, the price system does what policy can only awkwardly attempt. It is a communications network that no computer can emulate.

But prices do not just mediate exchange and production. They are sources of income and arbiters of what we get for the money we spend. They perform simultaneously the impersonal task of allocating resources and mediating the transformation of resources into consumable goods and services and the more divisive task of determining the sizes of individual incomes and how much those incomes will buy.

Disputes over energy prices arise in the tension between these two functions. The first function—allocating resources to their most valued uses, matching demand with supply, evening out shortfalls and surpluses, minimizing waste, and linking the present to the future through investment and choice of technology—involves common interests. But in the second function—determining, through the prices of personal services and farm products, the prices of labor and property, by region and industry, occupation and opportunity, ownership and training and skill, whose income will be large and whose will be small, and determining, by age and health and family size, by climate and location and consuming propensities, who gets more or less for their income when they spend it—conflict is inherent.

This second function, the one that divides us, is immediate and familiar. Anyone who drives knows what it means when the price of gasoline goes up: Drive less, or spend more on gasoline and have less for other things, or both. When the prices of electricity and heating fuels go up, purchasing power goes down. Higher prices always look like bad news to the consumer, and the bad news about energy prices comes across especially loud and clear. Gas, electricity, heating oil, and gasoline are standardized commodities whose prices are too conspicuous to go unnoticed.

At the same time, people whose livelihoods depend on coal, whether they work in the mines, own them, or make their livings in mining communities, are aware that booming coal prices are good news and sagging prices bad. Truckers see the cost of diesel fuel coming out of their earnings. And the most visible consequence of price control is that something costs less. These are firsthand experiences.

The impersonal role of prices is not so immediately apparent. Prices measure what natural gas or gasoline is worth to the people consuming it; they can also reflect the costs of drilling wells and building pipelines and operating service stations. And by signaling these values in a common currency, prices can indicate whether another gallon or cubic foot is worth more to someone who pays for it than it costs in the resources needed to produce it. The price system assembles, processes, and transmits this information continuously.

Furthermore, prices are the *incentives* that induce people to consume less of something that they value less than others do. Prices encourage the use or consumption of things that are worth more to a producer or consumer than what equivalent resources could earn in other uses. Scarcities arising from interruptions in supply or surges in demand generate the price increases that cause people to respond by consuming less, drawing on stocks, or increasing production. When gasoline prices rise or fall, people make their own decisions about their driving priorities. Anyone who would rather have the money that someone else is willing to pay for gasoline responds to his or her own priorities. And that decision, together with millions of decisions like it, is reflected *in* the price and is also a response *to* the price.

The average person does not need to know about gaps and shortfalls and surpluses. Prices signal what people need to know and stimulate their responses. The person who reroutes a truck of bottled gas to where the price is marginally higher or the homeowner who fills his oil tank in the summer to save a few cents a gallon does not have to know how many others are doing the same thing. The price system economizes information; a myriad of details get summarized in a price.

PRICES AS AN ECONOMIZING SYSTEM

Even when someone has recognized that prices are the signals and incentives that channel resources into industries and determine the output mix of the economy and consumer choices among available commodities, a further step is necessary to appreciate why words like *coordinate* and *efficient* are appropriate.

For example, a careful consumer will buy only the gasoline that is worth more than whatever else the same money could buy. If the price reflects the cost of the gasoline, gasoline will be used only up to where its cost is matched by its value. And the cost is the value to consumers of the other things that could have been produced with resources equivalent in value to the crude oil, the refinery operations, and the other inputs to the production and distribution of the gasoline. Producers can afford to expand gasoline production only as long as the gasoline they produce is worth more to the people who buy it than the value of the resources that go into its production; and those resources, in turn, are worth whatever consumers would pay for the alternative goods that the same resources could be used for.

This is why a price system, to the extent it is allowed to be (and can be) freely competitive, is an *economizing* system. The cost of energy is knowingly paid by the people who consume it. The value of energy savings accrues to the people who save energy. Investments in new production or new technology are profitable when the new oil, gas, coal, or nuclear or solar energy is worth enough to the ultimate consumer to cover the cost of production. Substitutes for fuel, such as insulation and alternative modes of transportation, materials that use less energy and lighter-weight objects that require less fuel to transport, will be attractive and profitable only if they save more than the cost of the resources, including energy, needed to produce them.

DEMAND, SUPPLY, AND ALLOCATION

The economizing achieved in an ideal price system can be divided into three components. One is the *demand* (or conservation) dimension. Higher or lower prices induce less consumption or more, more conservation or less, according to each consumer's scale of priorities. The readier people are to switch their consumption to something else, the less the price will have to rise.

Second is the *supply* dimension. Fuels, such as Alaskan oil, that cannot profitably be sold at a lower price because they cost too much to produce or transport become economically interesting to producers at a higher price that covers costs. Whoever can expand output at the least cost will be the first to do so as the price rises. The particular fuels most nearly competitive in cost will be the first developed when prices rise.

The third dimension is *allocation* among uses and users. As long as the price is free to respond to unmet demands or unsold supplies, it will settle at a point where those who buy the fuel are those to whom it is worth at least as much as it costs. The demands that go unfilled will be those for which the fuel is not worth as much as it is in other uses. In popular discussion, attention is given to the role of price in bringing forth additional supplies and inducing conservation. Less attention has been paid to the role of prices in allocating *existing* supplies, especially during emergencies. That role will be examined in detail in Chapter 6.

Expectations of future prices are crucial. Some supplies of new oil, gas, and coal take a decade or more to bring into production; synthetic and nuclear technologies need even longer. Conservation requires forward-looking modernization or replacement of plant and equipment. The prices to which much of today's behavior is a response, the prices that affect current investment in future supply, are the prices expected five, ten, or even twenty years from now. When consumers invest in heating systems or insulation or evaluate the gasoline mileage of new automobiles, they have to anticipate prices several years in the future. New fuels that come on the market in ten years will be a response to today's assumptions about what prices will be twenty years hence.

ALTERNATIVES TO MARKET PRICES AND THE MOTIVATIONS BEHIND THEM

Despite the widespread intrusion of government into pricing in the private sector, intrusion that is sometimes welcomed or even demanded by business, labor, farmers, consumers, or other groups, ours is an economy in which there is a presumption that where prices work, they are to be left alone. In certain areas, especially "natural monopolies" such as the retail distribution of gas or electricity, price competition between companies does not occur. Moreover, there are policy domains in which government inescapably affects or determines prices; examples are utility regulation, leasing of federal lands, and procurement of strategic stockpiles. But the presumption that ours is a market economy suggests that we begin by identifying the main reasons for having a policy on prices.

We can begin with energy pricing as an instrument of *energy policy* itself, which may sound almost like a definition. In fact, however, energy *pricing* is often used or proposed as an instrument, not for coping with energy problems, but for assisting the poor, controlling inflation, protecting the environment, changing residential patterns, or reducing highway fatalities. Nevertheless, pricing has been an instrument for household conservation, fuel switching by industry, research and development, commercialization of nuclear or solar energy, improved mileage, the stimulation of oil and gas supply, and the acquisition of petroleum stockpiles. A frequent justification for intervening in energy prices has been that other interventions, intentionally or not, were already distorting the price system. Subsidies for solar heating and insulation or for the industrial use of coal, controls on natural gas, mileage standards for automobiles, and taxes on gasoline have sometimes been considered justified by the policy of underpricing domestic crude oil or by current or past favoritism shown to competing technologies.

A most important policy area for the control of energy prices or their modification by taxes or subsidies has been *income and welfare*. The primary motivation behind the regulation of crude oil and natural gas prices during the past decade has been to protect the purchasing power of consumers and the vitality of particular industries that are especially dependent on energy. Several concerns are distinguishable here: concern about consumers in general and low-income consumers (for whom fuel is often an especially large expense), concern about the beneficiaries of rising energy prices (e.g., owners of oil and gas properties), and concern about energy-vulnerable industries. Concerns about these four groups have been behind virtually all controls on fuel prices at the wellhead or at retail, as well as on electricity.

Some powerful motives for controlling prices have been *macroeconomic:* inflation, exchange rates, the balance of payments, income, employment, and production. World oil prices have had a dramatic impact on the overall U.S. price level, and the need to control inflation has been a powerful argument for controlling prices or decontrolling them very gradually. The possible conflict between the benefits of market prices and their impact on inflation is an issue of vital importance and is the topic of Chapter 3 of this policy statement.

Price increases have been widely regarded as the *problem* rather than as a symptom or as part of the solution. There is a tendency to see increasing fuel prices as arbitrary transfers of purchasing power, not as reflections of genuine costs or responses to the need for conservation and enhanced supply.

Measures that hold down the price of imported crude oil can indeed protect our collective purchasing power. But in attempting to insulate ourselves from the real costs of energy, there is danger that we are deceiving ourselves, believing that if we do not pay the costs directly, they do not have to be paid at all. Price regulation can disguise the way fuel costs are paid and, to some extent, who pays them. It can sometimes redistribute costs, but it usually cannot reduce them. It can postpone the impact, but if it does, it also postpones the needed adaptation. Holding prices below costs cannot avoid subsidizing excessive consumption, inhibiting exploration and investment, and misrepresenting the worth of technologies that can economize energy.

Important corrective steps were taken by President Carter, who initiated phased deregulation of crude oil, and by President Reagan, who completed deregulation ahead of schedule. Natural gas is still regulated at the wellhead (although that regulation is being progressively reduced) and is subject to its traditional regulation at retail as a public utility. Most electric utility rates are determined or regulated by public bodies. Therefore, the deregulation of crude oil and refinery products did not dispose of the subject, and even the issue of price controls on liquid fuels could arise again suddenly in a severe oil emergency.

PRICES AS A SYSTEM

Controlling fuel prices entails something more than what happens when prices are tilted by subsidies and taxes, regulations and prohibitions. Consider gasoline. We can tax gasoline itself or the weight or horsepower of vehicles; we can enforce speed limits, provide fast lanes for car pools, and subsidize mass transit; we can encourage the four-day week to reduce commuting; we can impose or remove tariffs on foreign cars; we can relax or tighten emission standards on engines. In a multitude of ways, we alter the demand for gasoline and the supply and thereby affect the price, or we change the price directly with taxes and regulations that affect the cost of production. Still, distorted and intruded on, buffeted and manipulated, the price system is there. Gasoline does not go to those who arrive first or wait longest; it is not allocated by formula among a supplier's customers; it is not burned by somebody who would rather sell it but legally may not. It is available for those who will pay the price. (If not, the price will rise until it is.) Who gets the gasoline? Whoever chooses to buy it at the price. There is no need for another system. As long as prices can move in response to demand and supply, they balance out any discrepancies.

In contrast, price controls do not by themselves constitute a system. Controls *remove* precisely that element in the system that brought demand and supply into balance. When controls are imposed, some alternative mechanism will determine, by intent or by default, how much gasoline is produced and who buys it. In the absence of an administrative mechanism to make those determinations, something has to give; gasoline will be distributed by favoritism, waiting lines, black markets, or the sheer uncertainty that inhibits burning it today for fear that there will be none tomorrow.

The fact that government or private parties establish an allocative system does not mean that the system works. In principle, gasoline was allocated through a comprehensive system while prices were controlled during the gas lines of the 1970s. Supplies were allocated among locations and users but not among consumers. The result, in addition to a small overall shortage at controlled prices, was newsworthy local shortages from city to city and from city to countryside, as gasoline demand eluded formulas based on historical use or predicted need.

Natural gas under price regulation has not behaved so erratically for the homeowners. The consumer has a more stable relationship with the gas pipe than with the local gasoline pump; those who had gas in their homes usually had little need to worry, and those who could not get a gas connection had other fuels to worry about. But homeowners' satisfaction was due largely to the deliberate concentration of shortages and interruptions in business and employment. To evaluate direct controls, therefore, one has to evaluate the allocation system (or lack of it) that goes with the controls. Price controls do not operate in a vacuum. When prices are not free, something has to give. Until we have identified that "something," we have not identified the allocation system.

OPEC AND THE WORLD OIL MARKET

It is often argued that the idea of a competitive market is irrelevant to energy, especially petroleum, because the world market is dominated by the Organization of Petroleum Exporting Countries (OPEC), a cartel or political coalition that contradicts the very idea of a free market.

It is important, therefore, to distinguish between the domestic energy economy and the world energy economy. When we refer in this policy statement to markets and prices, we mean those markets over which the United States has jurisdiction and for which U.S. policy can be decisive. This statement is not about how the United States should approach OPEC negotiations or boycotts or embargos, trade with the Soviet Union, or economic relations with Iran or Mexico. It would, indeed, be a mistake to base policy toward the world oil market on the premise of competitive prices.

The United States shares with many oil-importing countries an interest in restraining our collective demand, particularly during supply interruptions and emergencies, to keep the price that we must pay to oil-producing countries from rising as steeply as it otherwise might. Recognizing that there is no world free market in petroleum is not in any way inconsistent with letting prices in the domestic economy play the role that is denied them in international trade.

APPROPRIATE ROLE FOR PRICING POLICY

On what grounds can government policy toward the pricing of energy properly be advocated? Under what circumstances do the arguments have merit?

MACROECONOMIC OBJECTIVES

Promoting efficiency is a *microeconomic* role. Prices accommodate supply and demand in individual markets. They allocate investment, inform technological choices, and stimulate supply or induce conservation at the level of individual decisions, commodity by commodity, industry by industry, fuel by fuel. Prices mediate the infinitely complex possibilities for substitution among fuels and among goods that differ in the energy used in their production; they are the connecting links between gasoline and heating oil, heating oil and natural gas, natural gas and electric heat, gas for heating and gas for making steam, electric motors and diesel engines.

But *macroeconomic* events are typically not priced within the system. The government budget and the money supply are superimposed on the price system. Incomes derive from decisions and actions mediated by the price system, but there is no market in which prices balance our diverse views about equality and inequality, the fraction of the population below some designated level of poverty, or the differences in income between old and young, or families with children and families without.

The dynamic wage-price processes that contribute to inflation are embedded in the price system; but unlike the prices of individual goods and services, which are always relative to other prices, the general price level lacks the self-equilibrating character of relative changes in individual prices. The rate of inflation, although generated through the price system, is not itself "priced" in the system as goods and services are.

For these reasons, income and employment, the balance of payments, the rate of inflation, and the distribution of income are bound to be issues of policy. The energy sector of the economy is so large and so susceptible to regulation that it invites efforts to regulate or manipulate it in the interest of combating inflation, strengthening the dollar, or protecting the purchasing power of the poor or the elderly. Of the five specific topics examined in this statement, two are macroeconomic: policy toward the poor and policy toward inflation.

NATURAL MONOPOLIES AND EXTERNALITIES

The price system can operate efficiently only under certain conditions. Competitive access to resources and markets is crucial. Certain industries, however, especially the retail distribution of electricity and natural gas, are technically incompatible with competitive duplication and have been regulated as public utilities. Few customers have any choice of where to buy their electricity or their gas; and when a franchise is granted to a sole provider, public regulation goes with it. Retail electricity rates are determined by public utility commissions and boards and government bodies in all fifty states and the District of Columbia. Wholesale deliveries of electricity and natural gas, because they are transmitted over long distances and cross state boundaries, are regulated by the federal government. There is, therefore, no escaping the need for policy in these two important energy sectors.

Another essential condition for an efficient price system is that all costs and values be reflected in the prices of commodities. But many products of economic activity are not captured in prices, and these externalities, some beneficial and some harmful, cannot be ignored. Much of the knowledge gained from research, for example, cannot be captured as a proprietary asset by whomever finances it. (This will be particularly true of experiments that test the environmental hazards associated with some new fuel. There can be little reward to the pioneer who discovers, for the benefit of all who are watching, that a particular new product or process will in the end be environmentally unsuccessful.) The development of nuclear power has been an obvious instance of this principle, but it applies equally to new ways of using coal or constructing energy-efficient buildings.

There is dispute about the nature and measure of the damage to heart and lungs, crops, forests, wildlife, painted surfaces, and countryside caused by the burning of fuels, and about how much and what kinds of abatement or compensation are worthwhile. But there is no question that many of these environmental impacts are outside the market and not reflected in the prices people pay for the energy they consume. Most of these environmental externalities have been internal in the United States because of the size of the country and its distance from other continents. But global changes in climate caused by atmospheric carbon dioxide from the burning of fossil fuels are now becoming recognized as a potential concern in the next century. Acid rain in Canada and Scandinavia is an international issue; the effects of other combustion products, as well as radioactive substances, will have to be treated as international problems.

GOVERNMENT AS A PARTICIPANT IN ENERGY MARKETS

Governments are themselves economic actors with influence on the market. Deposits offshore and on public lands are subject to leasing, licensing, military allocation, or other controls. The manner in which offshore rights are auctioned or licensed affects the pricing of fuel. Both the acquisition and the ultimate release of emergency reserves of crude oil, whether privately or governmentally owned, entail government decisions that affect prices, and so does a decision to use oil from military reserves. Moreover, the federal government is itself a purchaser of large amounts of fuel. Clearly the government is too large a participant in energy markets to be unaware of its influence on prices.

IMPERFECTIONS AND IMPEDIMENTS IN THE MARKET

Although prices are the signals to which the entire economy responds, not all those signals are received loud and clear. Not everyone knows how to read the signals. Not all markets work perfectly. Not all valuable innovations catch on. Not all opportunities for new conservation services promptly call forth new enterprise. Our basic commitment to the principle of market prices necessarily depends on the assumption that markets work well if not perfectly.

Some impediments to the efficient working of markets are institutional: regulations on mortgage loans for household investments in energy conservation or in new heating technologies, building codes, labor practices, landlord-tenant relations, traffic regulations, regulation of trucking and railroad and airline industries, difficulties of coupling industry with utilities in the cogeneration of electricity, and a multitude of other federal, state, and local constraints and practices. Many of these institutional constraints were innocuous until energy costs zoomed but now are significant barriers to the market's response to the prices of fuels.

In the endless dispute about how much consumers respond to prices, a judgment consistent with the recent evidence is that they respond more than skeptics expected but not as promptly as a theory of the rational consumer suggests. Besides understandable delays in turning over durable equipment such as cars and furnaces, there is sluggishness in consumer response. Procrastination, especially in the face of unfamiliar needs for new kinds of services for which no reliable local industry yet exists, has been aggravated by uncertainty about whether the energy problem is genuine and whether prices are going to stay elevated and rise even higher.

Some businesses have been sluggish, too. Builders and operators of commercial buildings and the small-scale local service industries (plumbers, oil merchants, house builders, home repairers, home-insulation contractors, air-conditioning contractors, and others) have been slow to acquire the skills and reliability to perform the home-energy services that homeowners and landlords cannot do themselves.

STRATEGIC CONSIDERATIONS IN OIL-PRICING POLICIES

Although the relationship between U.S. imports and OPEC prices is complex and unstable, the more oil we import, the more likely it is that OPEC prices will rise. The incremental cost to the economy of additional imports is greater than the nominal price because it pushes up the price on all imports.

This excess of incremental cost over average cost is even more serious because additional import demand on the part of any consuming country puts upward pressure on the OPEC price to all consuming countries.

There is also the problem of preparedness for emergencies. The United States and its allies are vulnerable to disruptions such as that in Iran, wars such as that between Iran and Iraq, political manipulation of prices as in the 1973–1974 period, Soviet activity, and even extreme economic strategies that might appeal to particular OPEC countries. Part of the cost of imported crude oil is the greater vulnerability that goes with imports. To the extent that it would be less if we imported less, this vulnerability can legitimately be counted as one of the incremental costs of imported oil, not reflected in the price but on top of the current-dollar outlay for the oil itself. The dangers here are both civilian and military, and they are dangers that we share with our allies.

Domestic pricing strategy in an oil emergency is bound to be controversial. Even persons convinced of the efficacy of the price system often doubt that prices should be relied on in an emergency in which liquid fuel supplies might be reduced by 15 or 20 percent or more. Indeed, it is widely believed that prices work well when people have time to adjust and the adjustments required are modest but that letting purchasing power determine who gets scarce gasoline or heating fuel in an emergency is bound to be disorderly and unfair.

This issue could become divisive quite suddenly. Decisions would be needed urgently. The worst problems might be avoided if thought through in advance. Rationing or allocating supplies or otherwise intervening to moderate the functioning of the price system would be a major regulatory undertaking. And whatever the United States chose to do would have to be coordinated with the actions of allied countries and consistent with treaty obligations. We devote Chapter 6 to price policy in an emergency.

SCOPE OF THIS POLICY STATEMENT

Five specific policy issues are examined in detail in this statement.

- Helping low-income families, whose household consumption of energy takes a disproportionate share of their disposable income, cope with the financial impact of higher fuel prices

- Accommodating energy policy to the battle against inflation

- Meeting the inescapable need for national policies on electricity prices

- Regulating production, transportation, and distribution of natural gas

- Preparing for emergencies involving the supply of imported oil

We begin our discussion with the treatment of low-income consumers because there is no more influential argument against a national policy of letting energy prices respond to costs and scarcities than the argument that to do so will injure the poor. Those who favor deregulation of oil and natural gas production may too easily be disregarding the impact that rising prices can have on poor families, especially families that live where home heating is a large part of their living costs.

We believe that controlling fuel prices is a poor way to cope with the energy problem. We also believe it is a poor way to cope with poverty. But this implies that there are other, better ways to protect the poor from bearing a disproportionate share of the burden of energy costs. And although the design of welfare programs to offset the impact of decontrol is not the function of this report, we have a responsibility to identify the magnitude of the problem. More than that, we have a responsibility to state our belief that government programs for the poorest part of the population have not adequately allowed for the burden of energy price increases, past or still to come.

The second most influential argument against deregulation has been that it would worsen inflation. There is no more urgent economic need than to bring inflation under control, to reduce and eliminate it as promptly as possible. Price deregulation has, inescapably, the immediate effect of raising the price index. The beneficial effects, on productivity and eventually on inflation itself, are less prompt, less obvious, and less calculable. An apparent choice arises, therefore, between pricing policies for energy and pricing policies for inflation. A satisfactory treatment of this subject is thus a prerequisite to recommendations for the pricing of energy.

One-third of our energy goes into the production of electricity, which is everywhere regulated in detail. There is no prospect of a competitive market in the distribution of this form of energy to final users; whatever the problems, deregulation of retail distribution is not the solution. Legally and institutionally, the complex problems of electric utility regulation are only partly accessible to federal influence. Utility regulation, therefore, promises no quick responses to policy initiatives. In our discussion, we have given it an emphasis commensurate with the size and permanence of the problem, not with the ease of implementing any solution.

Natural gas shares with electricity those public-utility features that make regulation at retail inevitable and inevitably difficult for federal policy to rationalize. But in exploration and production, natural gas shares characteristics with oil, including susceptibility to wellhead price control. And in its interstate transportation, it falls into a distinct federal regulatory class. In magnitude, the natural gas industry is the equal of the domestic oil industry. Much of what we say about electric rate regulation can be applied to gas and needs no detailed repetition; nevertheless, the discussion in Chapter 5 of natural gas contains a section that explores that fuel's complex role in our economy.

During the writing of this report, events in the Middle East and elsewhere reminded us of the urgent need to become prepared for oil emergencies. Unlike many who urge special preparations for energy emergencies, we conclude that primary reliance should be on prices, not allocations. The more severe the curtailment of supply, the more crucial the role of prices in avoiding the waste and misuse of precious fuel. But unlike some who envisage reliance on the price system, we find the macroeconomic consequences potentially disastrous unless there are advance preparations to cope with massive transfers of consumer income.

There is no doubt that other issues, from environmental protection and nuclear proliferation to synthetic fuels and solar technology, are equally deserving of attention. This discussion should show our appreciation of genuine reservations about leaving everything to the price system, just as it should make clear that although the proof may be forthcoming, the burden of proof should be on those who would make energy prices the target of government policy.

PERSPECTIVE ON THE FUTURE

A serious barrier to a rational response to energy prices will be the lingering effect of recent and current price controls and windfall taxes. These are bound to bias investment in new energy supplies or technologies. Decontrol of crude oil and gas prices remains controversial, and short of a constitutional amendment, there is little that any administration or Congress can do to guarantee to investors that the government will not step in again to impose price controls or punitive taxes. Prospective returns on new investments will, therefore, be biased downward, even if a free-market policy is currently pursued. Under these conditions, the fruits of free-market pricing will be less than they might otherwise have been.

The more promptly, completely, and unreservedly we deregulate, and the more we resist the temptation to capture windfalls, the more confidence we can begin to cultivate among investors that the rules of the game will not be changed to their disadvantage at the first opportunity or in the first emergency. Even consumers, both household and industrial, may be unwilling to protect themselves by making energy-conserving investments as long as they expect price controls to take care of the problem.

We have emphasized that the lead time for new fuels, new sources, new technologies, and new conservation measures is often a decade or more and that expected prices a decade from now will determine these decisions. Investors have to take their chances in a world of uncertainty. But if we want decisions taken now to incorporate the best estimates of what fuels will be *worth* in ten years, nothing is more important than building confidence that those future prices will be allowed to reflect the full values of those fuels.

CHAPTER TWO

ENERGY-PRICING POLICY AND LOW-INCOME CONSUMERS

We believe that increased reliance on the market system in the pricing of energy will bring supply, demand, and allocation of energy into better balance and will allow both consumers and producers to make intelligent, informed market decisions. But we also recognize that market pricing of energy will require consumers to pay the real cost of the energy they use and that under such a system, energy prices are likely to rise sharply.

In Chapter 4, we recommend that electric utilities be allowed to adopt rates that better reflect the replacement costs of electricity used; and in Chapter 5, we call for rapid decontrol of all natural gas. These steps are essential to assuring adequate supplies of both electricity and natural gas and to giving both business and individuals the necessary information on which to base their future consumption and conservation decisions.

Of course, all Americans will feel the impact of higher energy prices. But it is our strong conviction that no one group should be called upon to bear an unfair burden. That is why we believe that any national commitment to market pricing of energy must be accompanied by an equally strong commitment to making sure that the poor of this country, who spend proportionately more of their money on energy than any other income group, do not suffer in greater measure than other energy users.

Over the years, rising prices, including rising energy prices, have hurt low-income consumers more than any other group. Therefore, although this policy statement stresses the general benefits that will flow from energy price decontrol and utility rate reform, we urge that careful attention be given to the special problems that energy costs pose for those with few resources. Equity and decency require no less.

In principle, energy-pricing policy should be separate from welfare policy. The prices of few other goods or services are controlled to protect the poor. Market forces operate to set prices of such necessities as food, clothing, health care, and shelter, and a patchwork of welfare programs attempts to provide those living in poverty with cash or in-kind assistance so that they can acquire these necessities. But energy pricing is a different matter. For a combination of historical and practical reasons, it has been considered part of welfare policy. Whatever other reasons there are, which might include protecting middle-income consumers or providing certain industries with an advantage over foreign or domestic competition, protecting the poor is often cited as the principal reason for keeping energy prices controlled. Any short-term attractiveness of keeping energy prices down ought not to obscure the fact that such a policy has enormous consequences for the economy and for the well-being of all Americans, rich and poor alike. It is essential that this country bring its energy-consumption patterns into line with the underlying replacement costs of energy.

Having said this, we cannot go on to say simply that welfare will deal with poverty. To believe either that there is a coordinated welfare system in this country or that the various programs will automatically become more generous in the coming years is to be blind to plain facts. Political leaders, both national and state, now favor lowering welfare payments, not raising them, and restricting eligibility, not widening it. There is good reason to be skeptical of any solution to higher energy prices that depends on promises of new generosity from the welfare system.

Because energy prices have risen rapidly, and because energy accounts for such a large share of the poor's expenditures, low-income consumers have, on average, sustained a significant loss in purchasing power. But although higher energy prices have aggravated their problem, it is important to recognize that the poor have a welfare problem, not an energy problem. We calculate that the purchasing power lost to energy prices over the last decade amounts to more than $3 billion and perhaps as much as $7 billion *a year*. Moreover, the further price decontrols that we suggest later in this statement would mean that an even greater annual loss of purchasing power would be experienced in the future. Of course, any compensating changes in welfare would also have to take into account, along with many other issues, whatever changes rising (or falling) prices for other goods and services might require. Nevertheless, we are convinced that $5 billion or more would be a small price to pay for the benefits of getting energy prices right. Facing up to this welfare issue would be the best investment the nation can make if doing so would break the political deadlock over decontrol. Most important, any attempt to deny that a problem exists for the poor will inflict real misery on those least able to defend themselves against price increases and will invite a renewed deadlock on price regulation.

For many among us, this is a moral as well as an economic issue. We believe that there is an urgent need to decouple welfare issues from energy price control issues. But we also believe that we must not turn our backs on the need to allocate more money for welfare. The poor should not suffer more than the rest of society while the economic system adjusts to higher energy prices. We consider it imperative that the burdens of higher energy prices be shared more equitably than they now are and that the country give assistance to those who have no way to protect themselves. Those who are genuinely poor ought to receive enough money or direct aid to keep them even with the decrease in their real income caused by past and future energy price increases. However, protection against rising prices should extend only to those in real need, not to those more affluent who can respond to increases in energy prices by making changes in their consumption patterns. In that way, the nation's perennially limited welfare resources can be used to give meaningful aid where it is most urgently required, not to provide token benefits to all who ask. In the final analysis, we cannot expect the poor in this country to support a free-market pricing policy for energy unless the nation is willing to cushion them against the effects of those higher prices. Any other policy is unfair.

IDENTIFYING THE NEED

When the price of a commodity that consumes a noticeable portion of people's incomes rises faster than income as a whole, people have a natural tendency to think of themselves as being poorer. Something has to give in the budget. In the case of energy, this give needs to be encouraged because it provides the incentive for making protective investments in fuel-efficient cars, insulation, or more energy-conserving appliances. The give may also take the form of changes in our ideas of how much energy we really want and need as we discover that we want other things more than we want high-powered cars or 78-degree home heating in the winter. For all those who have some give in their budgets, any policy of keeping gasoline cheap, setting natural gas prices lower than their imported-oil heating-value equivalent, or holding electric rate increases below the cost of new capacity merely delays the inevitable and will only make that inevitable adjustment to the real costs of energy more painful. Each of us can see that we are not as well off as we once were, but public policy should encourage rational economic change, not create the illusion of stability. Unless the middle- and upper-income groups make the changes in their energy-consumption patterns that higher prices necessitate, there will be no halting the downward spiral that leads to greater dependence on foreign oil, decreased latitude in foreign policy, irreversible environmental damage, and perhaps even war.

Like consumers, businesses seek energy policies that work to their advantage. Even representatives of the Department of Defense and the General Services Administration have testified at utility rate hearings in favor of holding down the price of electricity used in federal facilities. We believe that there are no valid arguments for using uneconomically low energy prices to support any company or the government. Indeed, it is particularly important for those institutions that are large consumers of energy (and therefore especially sensitive to price increases) to face up to higher prices without delay. Higher energy prices ought to spur them to invest in conservation or change their product line. If higher energy prices force certain products off the market, that will only be the result of consumer preferences and market forces at work, as they should be.

The group of individuals who may justifiably be called poor and deserving of help in coping with rising energy costs is not so large that helping them is an impossible or prohibitively expensive task. Who are these people? For the purposes of this policy statement, the poor can be most simply defined as those who have no give in their budgets with which to adjust to rising energy prices and who therefore are forced to cut back on food, shelter, clothing, or medical care.

Between 12 and 20 percent of American families are economically disadvantaged. Each year, the Office of Management and Budget establishes a poverty line; eligibility for many assistance programs is set at 125 percent of that line, although some energy-assistance programs set eligibility at 150 percent or more. In 1981, 125 percent of the poverty level for a single person not living on a farm was $4,738; for a family of four, it was $9,313. In this statement, we use the commonly agreed-upon measure of 125 percent of the poverty level, which applies to about one-sixth of the population.

HOW DO THE POOR USE ENERGY?

Unfortunately, few studies have looked at the impact of higher energy prices on low-income families. The most recent data may be drawn from a U.S. Department of Energy study, the National Interim Energy Consumption Survey (NIECS), which investigated 4,000 households of various income levels.[1] Total energy expenditures rise with income level; that is, the

---

[1]/The results of the NIECS are contained in a number of Energy Information Administration reports and studies. Data cited in this paragraph are drawn from *Residential Energy Consumption Survey: Consumption and Expenditures, April 1978 through March 1979*, DOE/EIA-0207/5 (Washington, D.C.: Superintendent of Documents, July 1980).

higher-income groups use more energy per household than the poor do. But the rise is not proportional; low-income consumers spend a much larger share of their disposable income on energy than middle- and upper-class consumers do. For households with incomes between $3,000 and $5,000, fuel costs average 14 percent; but for households with incomes above $25,000, fuel costs average only about 3 percent. Assuming, for example, a 20 percent price increase and no change in consumption, the lower-income group would spend an additional 3 percent of its income for energy, whereas the higher-income household would have to pay about an additional one-half percent.

Data on the impact of higher energy prices on all fuel purchases are exceedingly difficult to assemble. Another study found that expenditures by the poor on natural gas and electricity more than doubled between 1973 and 1978, whereas average incomes for the group increased by only one-third during the same period.[2] A utility bill that required 6 percent of income in 1973 required about 9 percent in 1978. Moreover, these figures do not reflect the impact of rising prices of fuel oil (the most completely decontrolled residential heating fuel) and gasoline. Nevertheless, they give some indication of how higher fuel prices have affected the poor.

Low-income consumers are also hit by increases in the costs of energy used to make the goods and services they purchase. Although the effect of these "embedded" energy costs is still regressive, it tends to rise with income level, thus partially offsetting the more dramatically regressive effects of direct energy expenditures. The total cost of energy, direct and embedded, for low-income families amounted to about 22 percent of their income in 1974, and obviously it is higher today.

Even average costs are deceptive. Great variations exist among all user groups, including lower-income people. Fuel costs claim a higher proportion of household income in the North than in the Sunbelt because the poor who live in the colder climates are caught between higher energy costs and lower opportunities for employment in expanding industries. In all parts of the country, expenditures for natural gas and oil, both of which are used primarily for space and water heating, show less variation by family income than expenditures for electricity, whose versatility makes it easily used for high-value discretionary purposes. The size of the family, the background of the household, and the age of the head of household all influence energy

[2]/Harold Beebout, Pat Peabody, and Gerald Doyle, "The Distribution of Household Energy Expenditures and the Impact of High Prices" (Paper prepared for Conference on High Energy Costs: Assessing the Burden, organized by Resources for the Future and the Brookings Institution, Washington, D.C., October 9–10, 1980).

consumption. Some of the most tragic cases of hardship are to be found among those individuals who are helped inadequately (if at all) by welfare programs designed for the average poor person.

The tracking of energy expenditures and designing of assistance programs are further complicated by the fact that energy is delivered in a variety of forms and under different institutional arrangements. Gasoline is purchased from retail dealers; electricity and natural gas, from regulated utilities; and fuel oil, from independent distributors. Moreover, in the case of residential energy use, consumption may not be under individual control. Many users live in master-metered buildings, buildings where centrally generated heat and hot water are not controlled by individual thermostats, or buildings where maintenance is beyond the renters' control.

Energy consumption and its relationship to family income is a highly complex subject, and experts can argue about the details, many of which are quite important. Nevertheless, a clear message emerges from the available data: Because the poor must spend a greater portion of their income to purchase energy, an energy price increase will affect them more than it does a family with a more average income. Welfare benefits must be redesigned to take these relatively larger energy-consumption costs into account (at present, they do not). Until this is accomplished, assistance programs that do no more than keep up with averages reflected in the Consumer Price Index (CPI) will mean that the poor get poorer.

HOW MUCH ASSISTANCE IS NEEDED?

If we take the position that at least in the short run, the poor should be in no worse circumstances than they would be in if oil prices had risen only in line with the general price level, the cost of such assistance would be large, but not impossibly so. The best estimates suggest that poor households collectively had to spend about $7 billion more for energy in the winter of 1980–1981 than they did in the 1972–1973 period. Part of this increase can be attributed to the general inflation rate, but half or more represents the amount by which the poor are worse off in real terms because of higher energy prices. Thus, although the total is less than $7 billion a year, it is unlikely to be less than $3.5 billion. For the purposes of this discussion, $5 billion is a sufficient approximation.[3] These calculations were made prior to the oil decontrol speedup in January 1981; because of that action, the esti-

---

[3]/Alan L. Cohen and Kevin Hollenbeck, "Energy Assistance Schemes: Review, Evaluation, and Recommendations" (Paper prepared for Conference on High Energy Costs: Assessing the Burden, organized by Resources for the Future and the Brookings Institution, Washington, D.C., October 9–10, 1980, revised January 21, 1981).

mate is now a low one. Further price decontrol—for natural gas, for example—will push the figure higher yet.

Because energy must be purchased day after day, the additional $5 billion or more that we are discussing is a continuing expense for the poor and therefore must also be a permanent addition to the nation's welfare bill, adjusted upward for inflation and further increases in the real cost of energy.

A sense of scale concerning a proposed $5-billion annual addition to welfare outlays can be gained by comparing it with the size of the energy sector of the economy, which is now approximately $500 billion. In comparing these two figures, we are struck, not by how large $5 billion is, but by how small an expenditure it would take to offset completely the impact on the poor.

Those who speak, as we do, in favor of market-oriented pricing policies for energy have an obligation to point out how an expenditure that is comparatively small in the context of the energy sector could help the poor to cope with a market-oriented policy. Both the business and the conservation communities can and should address the genuine concerns of the poor at the same time that they advocate a true energy market.

PRIOR EXPERIENCE WITH ENERGY-ASSISTANCE WELFARE PROGRAMS

Experience with welfare programs addressed to helping the poor face rising energy costs is not encouraging. Programs have fallen far short of bridging even part of a $5-billion gap. In the first place, not all impoverished families have received aid; only about 60 percent of those eligible for energy assistance actually participated in the program in the winter of 1979–1980. If the same proportion had received aid during the 1980–1981 period, about $3 billion would have been required to put those who applied for assistance in the position they were in during the 1972–1973 winter because of higher energy prices ($5 billion × 60% = $3 billion). Yet, Congress appropriated only $1.6 billion for the winter of 1979–1980. In fact, the appropriation for energy assistance has never reached even $2 billion in any given year, let alone $3 or $5 billion, and it often has been passed months after winter's end. Other welfare benefits have not taken up the slack. Benefits under Aid to Families with Dependent Children have actually failed to keep pace with the CPI; and the Earned Income Tax Credit also fell behind for most recipients.

Nor have the programs designed to make the homes of the poor more energy-efficient been as successful as their supporters had hoped. Since 1974, the federal government has funded weatherization of the homes of the poor, and this program has been greatly expanded in the last several years. But although the pace of completion has picked up, only about 2 percent of the eligible families nationwide have received weatherization assistance, and funding for this program has now been substantially reduced.

Even the tools used to measure the amounts of assistance required have failed to take the special situation of the poor into account. As welfare is now structured in this country, a variety of different programs use independent measures of price changes to determine the level of payments from year to year. The CPI, the most widely used measure of overall inflation, has defects as an index for welfare payments, which ought to be based on the market basket of purchases made by the poor who live in different parts of the country under different circumstances. A series of more specialized indexes could provide a more adequate measure of how the poor are faring.

This statement does not attempt to deal with broad issues of welfare policy nor with questions of how programs to help the poor should be designed or administered. But our study has convinced us that no matter what old or new programs are used to deliver the resources, additional compensation ought to be channeled to the poor annually in order to cushion them against the loss of some $3 to $7 billion of purchasing power because of the energy price increases over the past decade.

IMPROVING CAPITAL GOODS

It is a characteristic of most welfare expenditures that they continue as long as the recipient is poor. For example, food assistance today does not prevent hunger tomorrow, nor does this week's rent supplement answer next week's need for shelter. Payments for energy conservation, however, are potentially quite different. This is one area in which good energy policy coincides with good welfare policy. By investing relatively small amounts in improving the homes and the energy-consuming appliances of the poor, government can meet twin goals. First, it can meet the welfare policy goal of reducing the need for ever-increasing amounts of energy assistance. Conservation is as good an investment when government is the housing owner as it is for any other owner who pays energy bills. Second, by making investments that reduce the energy consumption of the poor while providing them with the services they need, the nation as a whole benefits from reduced energy consumption. A barrel of oil not imported to serve a poor home is just as valuable as one not imported because cars are more energy-efficient. For this reason, we support government programs that provide capital assistance to the poor directly targeted to reducing their energy consumption.

One example of such aid would be an effective weatherization program. The existing program has suffered from considerable start-up difficulties. The program as originally conceived failed to deal with important categories of improvements. Until recently, for example, workers could not make structural repairs; consequently, ceilings could be insulated but broken windows could not be replaced. More importantly, no attention has been given to the role that more energy-efficient furnaces, water heaters, and refrigerators can play in saving energy. To the extent that current energy bills are paid for by welfare, capital expenditures for energy-consuming appliances need to be made on the basis of the same kind of rough cost-benefit calculations that an individual or an industry would use.

Weatherization programs are, of course, just common sense. But it is important that policy makers not forget that a small program of weatherization assistance is no substitute for the additional expenditures that need to be made to offset the real loss in income that the poor have suffered because of higher energy prices. There is simply no way to avoid the conclusion that additional welfare expenditures are justified by the facts.

CHAPTER THREE
ENERGY PRICES AND INFLATION

Inflation is the most threatening economic problem confronting the United States. Monetary, fiscal, and a variety of other forces have caused or aggravated inflation in this country; for the past eight years, however, rising fuel prices have been a major element in that inflation. Higher fuel costs not only take a direct toll on real income but also interact with other prices and wages to add momentum to the inflationary spiral.

Most of our recommendations would leave energy prices more free to rise in response to competing demands in order to get the energy problem under better control. Therefore, they risk making the economy more susceptible to inflation, at least in the short run. The urgent need to cope with the energy problem could thus conflict with the even more urgent need to brake inflation.

THE INFLATIONARY PROCESS

There is no straightforward accounting that determines, even retrospectively, the precise influence that energy prices exert on the level of prices in general. Inflation is not only a response to higher real costs of specific inputs; it is also a feedback process in which rising prices cause wages and other prices to rise, and rising wages cause prices and other wages to rise, in a circular and futile effort to catch up, keep up, or get a little ahead. When wages succeed in catching up in one sector of the economy, they push up the cost of production; at that point, somebody else's wages, salaries, or fees fall behind, and it is their turn to try to catch up. With fast productivity growth, everyone might keep up or even get a little ahead; with meager productivity growth, getting a little ahead of inflation becomes a competitive and somewhat futile exercise.

The numbers illustrate this. From 1970 to 1980, average hourly compensation in nonfarm businesses rose 118 percent, or 8 percent a year compounded. Whether hourly earnings were chasing prices or pushing them, or both, the average price level throughout the economy rose some 93 percent, or 6.8 percent compounded. The ratio of the index of earnings to the price index, 218/193, corresponds to the increase in output per hour (productivity) of private nonfarm-employed persons of 11.5 percent, or barely 1 percent a year. Correspondingly, unit labor costs for the same nonfarm business sector were up 96 percent, reflecting the ratio 218/111.5, the rise of hourly earnings over the rise of hourly output.

Of course, the numbers all fit; they have to. Prices were up, reflecting unit labor costs. Unit labor costs were up, reflecting wage and salary increases offset slightly by productivity improvement. Wages and salaries, on average, kept up with prices; but after a decade during which they more than doubled nominally, they were barely better in real terms than when the decade began.

Superimposed on this inflationary process were some nonsystemic shocks. Extreme weather was one, affecting food prices. The largest and most disruptive were the increases in the price of imported oil. In less than twenty-four months, from early 1979 to the end of 1980, the price of imported oil rose nearly 150 percent.

THE IMPACT OF OIL SHOCKS

Such an enormous rise in the cost of oil has had several distinct inflationary consequences.

The *first* impact—the transfer of income from American consumers to oil producers abroad—is readily measurable. On imports of 2.5 billion barrels a year, the price *increase* cost the nation about $60 billion a year. This is over 2 percent of the gross national product paid like a surtax to foreign oil producers, and it is a tax for which we receive nothing in return. We are getting the oil, of course, but we were *already* getting the oil and paying $40 billion for it at January 1979 prices. At the end of 1980, we were paying close to $100 billion a year for the same oil. The oil exporters are spending some of that in the United States, but they, not we, get the goods they buy.

That impact shows up as a direct pass-through in prices. This is where the consumer becomes aware that together we are $60 billion poorer. If the increase were paid as tribute out of taxes levied for the purpose, we would see it as lost income; but we actually see it as lost purchasing power, in the cost of the goods we buy. A large part of the increase shows up promptly in the direct cost of household use of gasoline and heating oil and some of it as an adjustment to the price of electricity. The rest shows up indirectly and less immediately in higher prices for food and other goods and services, reflecting the fuel and electricity used in their production, transportation, and distribution.

The total effect on the CPI depends on a *second* impact, the effect of *world* oil prices on the prices of *domestic* oil and gas (see Figure 1). In the absence of regulation, the price of domestic crude oil, for which imported oil can be substituted, would have risen with the costs of imports. And because oil and gas are close substitutes in many uses, the price of natural gas would have risen to an extent that is unpredictable (and controversial) but substantial. But price regulation (still in existence then for crude oil, although it was being phased out) postponed that rise in domestic oil prices. If we disregard this temporary postponement and do the arithmetic as though deregulation had been completed, the total increase in the cost of the oil resulting from the 1979–1980 price hike was about $140 billion annually. Three-fifths of that would have been collected by domestic producers and taxing agencies; two-fifths, by foreign producers. And had the price of gas

FIGURE 1

Consumer Price Indexes, 1955–1981

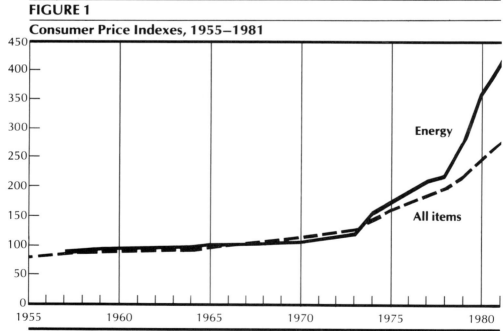

SOURCE: Executive Office of the President, Council of Economic Advisors, *Economic Report of the President* (Washington, D.C.: U.S. Government Printing Office, February, 1982), Table B-52, p. 291.

been permitted to move accordingly, the increase could have been another $50 to $75 billion, bringing the total to $200 billion or more.

We can now estimate the *direct* effect on the overall price level. Here, "direct" means the purely arithmetic calculation of what these cost increases do to the average prices of all consumer goods and services, without taking into account any responses of wages and salaries and other prices to this rise in the cost of living.

An approximation can be obtained by treating the price hikes simply as an addition to the money cost of the GNP. Using 1980 as the reference year, and recognizing that because of price regulation, only part of these increases was actually reflected in the 1980 GNP, we calculate an increment of roughly $200 billion to the money cost of a $2,500-billion GNP. That is 8 percent.

Energy is a larger component of personal consumption than of private investment and government procurement, mainly because of home heating and automobile fuel. The effect on consumer prices would, therefore, be larger than 8 percent. This is a one-time rise in the price *level,* brought about by an addition to the inflation *rate* over two years.

Part of this increase was delayed by controls on oil, which were finally removed in early 1981. The rise in the gas component will not be fully felt until gas decontrol is completed. But the increase is inescapable; it is built into the decontrol schedule.

THE WAGE-SALARY RESPONSE

We come now to the impact that cannot be estimated by arithmetic: the response of wages, salaries, and fees to the rise in the cost of living. It is this response that, after the oil shock has been absorbed, keeps the inflation rate going at a higher level. Some increases will be triggered by cost-of-living provisions in private and public wage-salary contracts; some will occur as a matter of corporate or institutional policy; some will be negotiated. Doctors, lawyers, and architects will raise their fees, impelled by the rising cost of living and the belief that their competitors are doing the same thing, that their clients expect it, and that their clients are doing what they can to keep up with inflation, too. Churches and universities are not exempt; they have to meet higher costs and salary claims and will raise their tuition and step up their appeals for contributions.

But this nearly universal effort to catch up with the escalating cost of living is not self-limiting. Some $60 billion a year of the lost purchasing power is simply going abroad. Any one among us can hope to keep up by raising fees or negotiating higher wages or salaries, but that pushes other prices up. Somebody pays the higher prices out of which those higher wages and salaries and fees come. And for anyone who gets ahead, somebody else falls behind. After obtaining wage and salary increases and higher fees, we are no closer to catching up with the escalating price level.

The domestic part of the increase, the rise in prices of domestic gas and oil, is not lost to the economy. A sizable portion of it goes directly to federal and state taxes. Corporations and individuals that enjoy higher incomes pay federal and state windfall and income taxes, whereas foreign producers do not. Some of the financial gains accrue to retirement funds and other institutional stockholders. The rest is a transfer to the owners of oil-producing properties, some of which will be plowed back into investment. The enhanced state and federal revenues can substitute for other taxes and so afford some relief. The effect on the CPI, however, is *not* eased by the fact that the higher proceeds on domestic fuels are not lost to foreign producers.

If there were no feedback, no response to the rise in the cost of living, the price *level* would simply be higher in perpetuity by that calculated amount, having got there by a transient (two-year) higher rate of increase. The impact of oil prices after the adjustment would be a step upward superimposed on the core rate of inflation. But that underlying rate embodies the upward pressures that are responses to the rise in the cost of living. As real wages and salaries fall back several more percentage points, the shortfall is certain to add to the core rate.

Thus, although it is difficult to disentangle from the other determinants of inflation, the *cumulative* effect of that oil shock is substantially larger than the increased oil expenditure itself because of the enhanced underlying inflation rate.

In effect, the productivity of the U.S. economy is reduced by 2 or 2½ percent because we pay an extra 2 or 2½ percent of our GNP for the oil that we import. (Literally, it is the "productivity" with which we convert exports into the oil we get in return.) And in addition to this real permanent loss, inflationary pressure is intensified by everyone's doing the only thing he or she can do to try to restore that lost purchasing power, seeking a larger share of the reduced output through higher wages, salaries, fees, or profits. The consequence is the familiar futile spiral.

To identify a rise in the price of imported oil as an inflationary pressure does not, of course, imply that the entire impact has to be taken in the form of inflation. Fiscal and monetary policy can be tightened further to contain the impact. Some of the cost will then take the form of lost production, income, and employment rather than greater inflation. Whatever the means chosen to combat inflation, and however rigorously they are applied to the problem, that *problem* is so much the larger because of the rise in oil prices.

Coming on top of a decline in productivity growth and in the midst of an inflation that has been gaining alarming momentum since the 1960s, the impact of oil prices on inflation has to be counted among the most severe effects of the entire energy problem.

INSULATING THE ECONOMY AGAINST OIL PRICE SHOCKS

As long as the United States remains dependent on foreign sources for a large part of its liquid fuel, it will have to give up goods and services in exchange for the imported oil. There is no way to avoid paying for it. The reaction within the economy, however, is a collective competitive effort to recover, through higher wages and salaries and the price increases that finance them, a loss that cannot be recovered. A further tightening of fiscal and monetary policies can reduce demand and somewhat inhibit wage increases, but the effect on labor costs may not be large or prompt enough to counteract the adverse effect on employment and production. Are there any mechanisms that can dampen, disengage, or insulate that domestic response and hold the price increases to the direct out-of-pocket cost of imported oil, quenching those secondary repercussions?

Price controls have been tried. They could not insulate us against the impact (including the inflationary-spiral repercussions) of the *foreign* price increases, but they did postpone the larger increases in expenditures for *domestic* oil and gas. The mechanism was to average the controlled domestic price of oil and the uncontrolled price of imports through an "entitlements" system and, separately, to control the price of gas. That policy was ultimately abandoned, though only very gradually in the case of gas (the decontrol of which is only partly accomplished); controls were deemed to do more harm for energy (and ultimately the cost of living) than good for the cost of living in the short run. We support that judgment. The same criterion would apply to any other technique that might disguise either the true cost of imported oil or the true worth to consumers of domestic oil and gas and the true cost of enlarged domestic production.

Two issues nevertheless remain. Can something be done about the price indexes that play such a direct part in triggering contractual and statutory increases in wages and salaries? And is *postponing* price increases and their inflationary consequences a legitimate role for controls?

DECOUPLING THE CONSUMER PRICE INDEX

There have been proposals to partially decouple the CPI from several kinds of income payments that are subject to statutory or contractual escalation. Whatever wisdom there is in such proposals needs to be judged in a broader context than that of energy prices. It would be a mistake, in both diagnosis and in policy, to attempt any attack on the indexing of public- or private-sector income payments—whether current compensation, retirement benefits, welfare payments, or property tax abatement—as though the problem were peculiar to energy. Just as we see no persuasive reason to protect particular groups from rising energy costs in a way that is different from the way we would want to protect them from higher food or medical costs, we see no reason to go in the opposite direction and detach cost-of-living adjustments from rising energy costs while leaving such adjustments intact for other components of the cost of living. Any campaign to change the institutional arrangements that have developed in this country for the indexing of compensation, retirement, welfare, veterans benefits, or any other form of regular payments should be a matter of broad and permanent policy, not fragmented into energy and other components.

There are, nevertheless, legitimate procedures that could be used to keep energy prices from distorting the official measures of the prices consumers pay. In an oil emergency, gasoline may be rationed, or an emergency tax-rebate system for gasoline or for all oil products may be put into effect. (These mechanisms are discussed in detail in Chapter 6.) In the event of rationing, it is essential that the coupons (or whatever physical form the entitlements to a gasoline quota may take) be transferable, exchangeable, and marketable. Such transactions in coupons should not be merely permitted; they should be encouraged. The result would be a consumer's price of gasoline made up of two components: a pump price paid in cash and a coupon price paid in coupons worth cash. The cost of gasoline to any consumer, whether he or she consumed the exact ration or less or more, would be the pump price *plus* the coupon value. That sum would be the cash forgone for every gallon consumed. The total cost of gasoline to all consumers, however, would be only the pump price. Because sales and purchases of coupons among consumers cancel each other out, the net expenditure on gasoline would be only the pump price.

The crucial point in the management of inflation is that the relevant price indexes should reflect only the money paid for gasoline, not the transfers among consumers who trade coupons. The coupon price should not be added to the pump price in constructing the CPI; or, if it is included, the money value of the coupons distributed should be treated as a negative gasoline tax, a price offset, or a cost-of-living bonus (the name does not matter) to make it clear that the coupon part of the price was fully offset by the distribution of free coupons. (There is, of course, a loss in real value that this treatment of the CPI is not intended to disguise; people are driving less. The proposed treatment of prices in the index relates only to the cost of the driving that people do, not to the hardship resulting from the driving they do not do.)

The alternative is an emergency tax on gasoline (or on all oil) with the proceeds promptly rebated. The merits of alternative rebate systems are discussed in Chapter 6; our point here concerns inflation. If the emergency tax is included in the CPI, the index should include the rebate as an offset. To the extent that the proceeds are rebated, there is no net loss of purchasing power for consumers as a whole. It would be a distortion, and an inflationary one, to reflect only the tax half of the two-way process.

A STRATEGY OF POSTPONEMENT?

When world oil prices rise, holding domestic prices of gas or oil at current levels can only postpone the ultimate inflationary impact. Some past inflationary shocks have been imbedded in the economy for later release through natural gas price deregulation. Consequently, the deferred shock, when it comes, will appear to result from deregulation rather than from the original increase in the world price.

As indicated earlier, we believe that regulating domestic prices in perpetuity would do more harm with respect to energy supply and use than it could possibly do good with respect to inflation. We supported the deregulation of both crude oil and natural gas prices.

But because regulation can temporarily insulate the economy from some of the repercussions of world price increases, it can stretch out or postpone the domestic impact. Does regulation have a constructive role to play in smoothing out the shocks by "scheduling" inflationary impacts over time?

There are at least three ways in which regulation could change the time profile of an inflationary shock. One is straight postponement. If it were always better to postpone the inevitable shock, if a future discount factor always applied, price regulation could put off the impact until next year or the year after.

A second possibility is that shocks do their worst mischief when they come in spurts. Containing the shock and releasing it gradually could translate jerky and unexpected increases into smooth and more predictable patterns.

A third possibility is that inflationary pressures alternately intensify and ease and that added pressures do more harm when existing pressures are already at a peak. Regulation could be imposed when pressures are severe but expected to ease in the future and lifted when pressures ease or are less than had been foreseen.

In each case, holding back the increases delays the impact on energy markets. As a consequence, the stimulus to conservation is postponed, the impetus for technological change is pushed into the future, and the rise in the return on investments in new and expanded energy supplies is put off. These postponements will add to the real future costs of energy.

As far as long-lead-time investments in fuel-consuming equipment or new energy supplies are concerned, foresight may lead to anticipatory decisions, and the delay in incentives may therefore not matter. But postponable decisions will be postponed. And delaying all those signals that require the intermediation of the market, that cannot simply be read by consumers and producers from current oil or gas statistics, will delay the responses.

So, clearly, there are costs in terms of energy consequences in pushing the impact forward in time. But are there counterinflationary benefits, and do they outweigh the costs?

We believe the answer is no.

First, it is far from certain, and not even especially plausible, that routinely putting off the day of reckoning makes added inflation easier to manage when it comes.

Second, it is also not certain, nor even especially plausible, that the ultimate cumulative inflationary impact is lessened by making it gradual rather than sharp.

Third, there is some plausibility to the idea that scraping the tops off the inflationary peaks and filling up the valleys would be of some value if fluctuations could be predicted with confidence. But nothing in our recent history suggests that inflation is so predictable. (Indeed, if we understood inflation well enough to predict it in that fashion, we would probably have it under control, and the issue would not arise.)

There is, furthermore, a temptation to manipulate prices, especially through postponement of an increase, for reasons other than minimizing the cumulative inflationary impact. Regulation of gas and oil has proved highly susceptible to political pressures and to interregional and other conflicts of interest and subject to manipulataion for too many purposes besides the management of inflation.

For all these reasons, we are persuaded there would be no net benefits from any policy that attempted, through price regulations of gas and oil, to contain the immediate impact of higher energy costs and release it in a strategy of planned counterinflationary deregulation. We doubt whether such a strategy is good anti-inflation policy. We know that it is bad energy policy.

CHAPTER FOUR
ELECTRICITY-PRICING POLICY

Electricity is a remarkably versatile form of energy. It can be used almost everywhere that other energy forms can, and there are many kinds of work that only electricity can perform. However, electricity is not itself a primary energy source; rather, it must be produced from sources such as coal, oil, uranium, falling water, natural gas, or sunlight, and that production takes place in generating stations that are mostly large and very expensive to build and run (see Figure 2). The production facilities are themselves predominantly owned by companies that have a franchise to generate and distribute electricity in a particular area.

Electricity generation consumes one-third of the nation's energy, including more than three-quarters of the nation's coal production, a domestic resource that is in abundant supply (see Figures 3 and 4). Electricity gen-

FIGURE 2
Electrical Generation by Fuel Type, October 1981

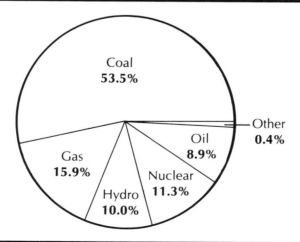

SOURCE: U.S. Department of Energy, Energy Information Administration, *Electric Power Monthly, October 1981* (Washington, D.C.: U.S. Government Printing Office, February, 1982).

34

eration is also a prodigious consumer of capital, accounting for about one-third of the funds raised by business in the domestic capital markets each year, and is perhaps the most capital-intensive industry in the world, whether measured by investment per annual dollar of revenue, by investment per employee, or in some other way.

Because the electricity industry is such a large user of this country's primary energy resources, we will have declared energy policy defeat if we flinch from making sure that electricity prices reflect the true costs of those resources. Throughout this statement, we stress the benefits that accrue to consumers and to the country from using market mechanisms. A smoothly operating energy market lets buyers know how much the next unit of energy will cost, so that they can use only as much energy as they really care to buy, and lets suppliers of energy know how much to produce in order to make an acceptable profit. In theory, these principles of pricing ought to apply to the sale of electricity. The price that all consumers, whether residential, industrial, or commercial, pay for every kilowatt-hour of electricity consumed should be the price of producing the last kilowatt-hour. Every consumer in a given utility system ought to pay the same price for electricity consumed at a certain time, after appropriate adjustments for differences in the cost of ser-

---

FIGURE 3

Primary Energy Consumption, by End-Use Sector, 1981 (Quadrillion Btu)

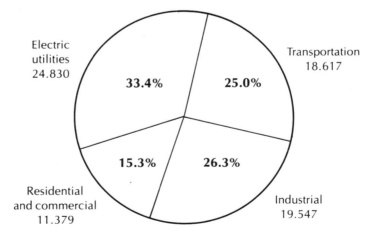

Electric utilities 24.830 — 33.4%

Transportation 18.617 — 25.0%

Residential and commercial 11.379 — 15.3%

Industrial 19.547 — 26.3%

Total consumption: 74.417 quadrillion Btu

SOURCE: U.S. Department of Energy, Energy Information Administration, *Monthly Energy Review* (Washington, D.C.: U.S. Government Printing Office, February, 1982), p. 21.

FIGURE 4
Electric Utility Electricity-Flow Diagram, 1981 (Billion Kilowatt-Hours)

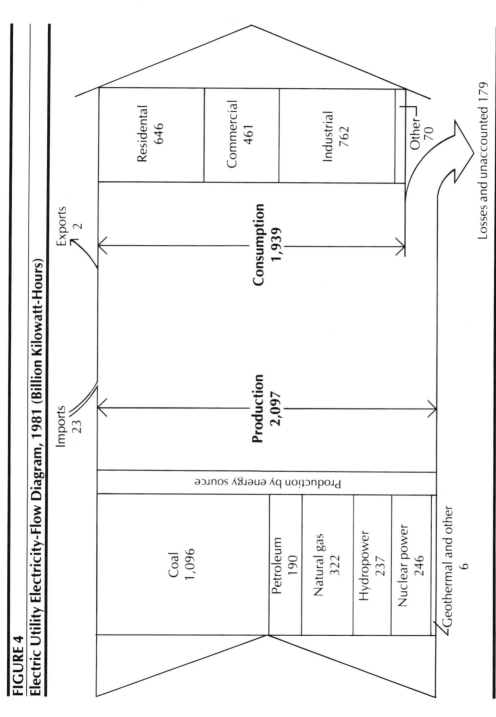

SOURCE: U.S. Department of Energy, Energy Information Administration, *Monthly Energy Review* (Washington, D.C.: U.S. Government Printing Office, February, 1982), pp. 10, 65, 66.

vice (which can be very large in some cases) is reflected. Each customer would then be able to decide whether more electricity had the highest call on the income available and would have the price information that would show whether the purchase of more efficient (but more expensive) energy-consuming equipment was worthwhile.

Every user of energy should know whether there are less expensive ways to accomplish the job being done. If information, whether it is provided explicitly or through the price system, is to be effective, it must be available at the time when the decision maker has an opportunity to choose a course of action. It is only at that moment of choice that the entire array of information, desires, resource availability, and alternatives can make a difference. Of course, at any time, we never really know which choice is the first in a series and which is the last. The only proper way to analyze decision making is to recognize that *every* decision is potentially made at the moment that the decision maker will cease doing one thing and choose some other. One of the most valuable pieces of information available to any consumer making a choice is the price of a commodity. Knowing the price, which ought to represent the resource cost that society devotes to creating and delivering the good, the decision maker can then examine his or her desires, alternatives, and resources to see which choice is truly best.

Of course, many decisions are relatively short term and trivial, even if they involve energy consumption. But price information subtly influences even small choices such as turning off lights and setting back thermostats at night, exerting influence without government regulation or intrusion. Larger decisions, such as the purchase of a particular model of air conditioner or the adoption of a new industrial process, are more strongly influenced by energy prices. In such cases, the expenditures and potential savings may be large enough to justify taking the time to calculate the total cost of owning and operating the equipment over its useful life. Knowing the long-term availability of energy is also an important factor, of course. But every decision, small or large, will be made more intelligently if the price accurately reflects the resource cost to society.

If the theory is so straightforward and conventional, why does the electric-utility industry not simply price its product according to these principles so that the country can benefit from the efficiency gains that would follow? Part of the answer lies in history and politics. But some genuine difficulties lie in the way of pricing all electricity at its replacement cost. Those difficulties are serious enough to lead some observers to believe that we may never be able to price electricity according to free-market principles and that the best that can be done is to make a series of compromises.

THE HOW AND WHY OF ELECTRICITY PRICING

The process of setting electricity rates strives to assure that the embedded costs of generating, transmitting, and distributing electricity are paid for each year and that customers pay for electricity at rates that bear a reasonable relationship to the costs they impose on the system. If the rate setting is done properly, the total amount of money that all customers pay to the utility over the full year ought to equal the amount of money the utility needs to conduct its business. Because the rate setting is done for anticipated future sales and costs but based on past experience, the revenue level may fall short if sales drop or expenses rise relative to expectations. Even annual rate adjustments may not be frequent enough to track rapid changes.

Public regulation of rates and conditions of service is well established. Early in the history of the industry, it became clear that if a large number of companies distributed electricity in the same area, there would be duplication of facilities that are very expensive to construct. Therefore, as in the case of the railroads, the telephone companies, and the water companies, electric companies were granted exclusive franchises to sell electricity in particular areas. In exchange for this protection from competition, electric utilities gave up their freedom to set their own prices and accepted the obligation to provide adequate service to all customers without unreasonable discrimination.

The public control of electricity rates also served to insulate both the public and the companies from rates that mirrored those a market might have produced. Consumers, with no choice of companies, were protected from overcharging by the utility company because the public-utility commissions had to approve rates before they could be collected. The companies, on the other hand, were able to charge more for electricity in some periods (notably in the first two post-World War II decades) than economic theory would have justified, given the fact that the costs of new production declined significantly.

CHANGED CIRCUMSTANCES

Circumstances have changed drastically in the last fifteen years. Electric utilities are now confronted with the fact that (except where oil-burning facilities are being replaced by those using fuels sufficiently less expensive to offset higher capital costs) new units of production no longer bring cost reductions to the customer or the utility. The efficiency improvements that used to flow from new power plants are generally a thing of the past. Today, most new investments in expanded capacity do not lower average costs;

they raise them. Compounding the damage caused by the fact that economy-of-scale limits using present technology seem to have been reached is the fact that costs of simply replacing existing plants have risen faster than the general rate of inflation. Thus, electric utilities are in an even worse position than industry as a whole, in that their regulated depreciation charges understate their current replacement costs by more than just the overall inflation rate.

The current system of setting rates makes the costliness of new or replacement generating plants an important issue for utilities because they must average their historical costs in setting their rates, rather than basing their rates on today's costs. Utility commissions do not allow rates based on the replacement costs of the older plants; consequently, electricity generated from these plants is priced with unrealistically low interest and depreciation rates as a component of the cost. For these reasons, no customer pays the true replacement cost of electricity (except perhaps rarely and by accident).

There is an additional reason why electric rates fail to equal replacement costs. Cost averaging is applied not only among plants built at different times, but also to plants with different operating costs over a given day or year. There is another reason for cost-averaging rates. Because electricity must be generated just as it is used (storage in meaningful quantities is unacceptably expensive), the cost of generating electricity varies significantly from season to season and from hour to hour. Utilities operate according to a principle of "economic dispatch": Lower-operating-cost units are put on line first; As demand rises during the day, more expensive units are switched on; The most expensive "peakers" come on for only a relatively few hours a year. Because electricity is typically sold to residential and smaller commercial and industrial customers for the same price at every hour of the year, those users during the peaks get a bargain subsidized by other users who make demands during off-peak hours.

These are two rather different problems, one arising out of the way regulators permit rates to be set, the other out of technical factors. To solve either problem, a utility would need to collect detailed cost data and then use those data as the basis for a rate structure. The detailed cost data include an accurate calculation of the replacement cost of electricity at every hour during the year, using past experience and forward planning as guides. These costs can vary enormously, ranging from perhaps 3 or 4 cents per kilowatt-hour for off-peak electricity (i.e., that generated at a time when there is plenty of reserve capacity in the system) to 12 to 15 cents or even more at times when the most expensive units are on line and there is little or no reserve left. Utilities throughout the country are now undertaking or have completed such cost studies; the Public Utility Regulatory Policies Act (PURPA), which is part of the National Energy Act of 1978, requires utilities of all kinds to perform studies of time-differentiated costs as part of a larger series of mandated state hearings on rate-making practices of gas and electric companies.

Once the bookkeeping work of calculating these costs is finished—a task requiring far more sophistication and judgment than those who have not participated in the process might expect—the costs would have to be translated into a rate structure that approximated the relevant costs.

The electric utility and its public regulatory commission could design such rates in one of several ways. Time-differentiated rates are a clear first step because the differences that exist in the cost of generation between peak and off-peak hours are both large and easy to understand. Like long-distance telephone rates, which alert users to those times when use of the network and switching equipment is relatively expensive and when it is cheap, time-of-use electric rates would alert customers to when they could use electricity comparatively cheaply and when it would cost more. Again like telephone rates, electric rates based on time of use ought to be simple enough so that users can easily understand in advance what the costs are likely to be. Although residential electric meters to accomplish this time-of-use recording are now rare and therefore appreciably more expensive than ordinary kilowatt-hour meters, technological advances and a large market for them could make such meters cost-competitive.

Unfortunately, the larger problem caused by the fact that rates are based on the old, historical cost of generating plants rather than on today's costs cannot be solved so easily. Virtually every state public utility commission believes that electric utilities should not be permitted to charge the replacement cost for electricity generated by a low-cost plant that was built years ago. Instead, the commissions require that all historically based accounting costs carried on the books be homogenized to achieve the average cost at which the electricity is to be sold. But today these historical costs are nearly always dramatically lower than the replacement cost because they do not take inflation and real cost increases into account. For example, the depreciation allowance that can be charged for a plant as a component of current costs is inadequate to replace the plant at current costs. Pricing is further complicated by the difference between what many commissions say they will allow their utilities to earn and what they actually permit as a result of their rate decisions. The effect of decisions that hold down rates of return is to make borrowing and equity financing more expensive for the companies and, perhaps, to slow down cost-effective investments.

If the utility commissions were to permit the electric companies to base their rates on the present-day costs of plants, spread over the kilowatt-hours sold, the practical effect would be that the companies would collect far more money. In many cases, the extra revenue generated would be substantial. The most dramatic example of this would be in the Pacific Northwest, where the average electricity price per kilowatt-hour may be about 3 cents to the residential consumer; but if on-peak electricity comes from newly installed nuclear capacity, the cost to the customer would be five times as much. Because *all* electricity sold to all consumers during a relevant time period would carry the higher price, the extra revenue collection would amount to about 12 cents per kilowatt-hour sold. But even a more typical midwestern utility system illustrates the point as well. Residential electricity rates are about 7 cents per kilowatt-hour now; electricity from new capacity would cost 10½ cents per kilowatt-hour, a 50 percent increase in price. In systems that already collect billions of dollars annually, such increases would amount to very large sums.

The increased revenue that this pricing reform would cause is the exact analogue of the problem that resulted from decontrol of crude oil prices. In the case of crude oil, there were two sources of supply: one that was inexpensive because of price controls (domestic oil), and one that was priced at the replacement cost (world oil). Before decontrol, the prices of the two were averaged by means of the crude-oil allocation system, and all oil was sold in this country at an average price that was below its world price. At the time of decontrol, the additional revenues that flowed to the domestic producers of oil were partially taxed away by means of the windfall profits tax (actually an incremental *ad valorem* excise tax unrelated to profits). The proceeds of that tax were used for a variety of purposes.

The nature and history of the institutions that control public utility pricing are such that a similar solution for electricity pricing would represent a major shift in attitude. If such a system were in place, utilities could be ordered to invest in the metering necessary to implement pricing changes, to collect the revenue that pricing at replacement costs generated, and to have any revenues above the expenses (including more realistic depreciation reflecting cost inflation) and adequate rate of return taxed away.

To be sure, a tax can be as subject to the unhealthier meddlings of the political process as price regulation is. Because the tax would be new, there would be debate about how to set it correctly and temptations to make it either larger or smaller than economic theory would dictate. Administering a new tax can present difficulties. The very size of the extra revenues that such a tax might raise would make it an attractive target for use in financing any number of special or favored purposes.

Notwithstanding these concerns, however, such a system could work to make electric utility rates contribute to the efficiency of the economy's production and consumption of electricity while preventing any great transfer of wealth to the franchised utilities.

EXPERIMENTS TO MAKE THE EXISTING RATE STRUCTURE MORE RATIONAL

Such a substantial change would take time to debate, refine, and implement. Meanwhile, utility commissions intent on trying to make the pricing system more rational must seek methods of pricing, regulation, or combinations of the two that move in the direction of the results a more rational system would create. A number of experiments and approximations have already been attempted. However, attempts have often been flawed in either theory or execution or both.

INVERTED RATES

In the past, rate structures charged a great deal for the first few kilowatt-hours consumed and progressively less for larger blocks of consumption. A typical rate structure might have charged $5.00 for the first 10 units, 6 ½ cents per unit for the next 90 units, 4 ½ cents per unit for the next 100 units, and so on. This kind of structure was defensible as long as new-production costs for electricity were falling and added consumption meant savings for all consumers; but under today's conditions, it no longer makes sense. One rate structure that has been suggested is to lower the unit cost to those who consume less electricity and raise it for those who consume more. If done correctly, proponents argue, the inversion would convey the general information about rising costs to consumers. Moreover, because the undercharging at the lower end of the scale would exactly balance the overcharging at the upper end, the utility would not collect more revenue than it now does. Unfortunately, the system would bear only an accidental relationship to rates that would be justifiable according to economic principles. No customers (or perhaps only customers with very large consumption) would pay replacement costs for all the electricity they used; most customers would still be shielded from the information that would enable them to make accurate consumption and conservation decisions.

LIFELINE RATES

Lifeline rates, a version of inverted rates, have been instituted in a number of states and rejected in others. The purpose of lifeline rates is to subsidize purchases by customers who use electricity in small quantities. As in the case of inverted rates generally, the subsidy is funded by sales at above-cost rates to larger users. PURPA required that all states hold hearings to consider adopting such rates. California has instituted rates that take climate into account and that provide sharp incentives for large users to lower their electricity consumption. Lifeline rates are based on the observations that poorer people tend to use less electricity than richer people (although the correlation is far from exact and in many cases does not exist at all) and that a certain minimum amount of electrical use is virtually nondiscretionary in this society. Lifeline rates, whatever their other goals, fail to approximate rates set according to economic principles, mix welfare policy into energy-pricing policy, and raise opportunities for unwisely subsidizing electricity use.

INVESTMENTS ON THE CUSTOMER SIDE OF THE METER

Quite a different solution is provided by breaking the traditional barrier between the electric company's side of the meter and the customer's side. A utility company that is faced with deciding whether to build a new plant confronts the total present-day cost of new capacity directly. But with average-cost pricing, the consumer who purchases electricity evaluates every decision by comparing the apparently low-cost electricity with the present-day-priced conservation investment. Traditionally, the electric meter has operated as an absolute barrier between customer and company, but some companies are now breaking that barrier and making efficiency improvements on the customer's premises. These improvements, which both save electricity and help put off construction of new capacity, are then included in the rate base as if they were ordinary generating or distribution equipment. (Some companies treat the investments as zero-interest loans that must be paid off when a home is sold.)

However, the existing conservation-investment programs have three features that make them less than ideal substitutes for more substantial rate reform. First, they are generally limited to homeowners and therefore provide benefits to only a portion of the customer base. Second, they are limited to providing only the most conventional efficiency-improving equipment and materials, such as insulation. These limitations exclude many more dramatic savings, such as those that could be gained by replacing old appliances, air conditioners, or furnaces. Third, in the interest of equitable treatment for those customers who have paid for insulation themselves, the utilities have limited their programs to installing only an amount of insulation that will lower the cost of new generation for all consumers, both those who participate in the program and those who do not. The amount of insulation that meets this equity test may actually be less than what would be a justified investment for the utility itself if it could choose between the cost of building new plants and the cost of avoiding such construction.

The fundamental difficulty with this system is that it can provide only capital equipment changes for consumers; it cannot exert cost pressures to motivate them to change the ways in which they operate their homes or businesses. Even if this limitation is accepted, the system has another flaw. As a long-term substitute for proper pricing, it holds little promise unless the country is truly willing to let electric-utility companies make investments in any kind of electricity-consuming sector, from new aluminum-production facilities to computer-controlled commercial office heating-cooling systems. Because present rate regulation fails to give proper price signals to customers, a substitute that permits the utility to make conservation investments ought not to be limited to a few items; once the barrier is broken, it is hard to see why utilities should not be permitted to make *any* investment that would avoid more costly new construction. As a practical matter, however, few (if any) utilities have the funds available to make such investments under current conditions.

LOAD MANAGEMENT

Another nonrate strategy is remote operation of appliances located on customers' premises in order to shave off the top of the peak and reduce the pressure to build new plants to meet peak-capacity demands. Appliances and industrial processes that can be interrupted without noticeable effect (such as water heaters or air conditioners in which hot water in the tank or air in the already-cooled room reduces the impact of short interruptions) are wired to respond to signals transmitted over the electric wires or via radio or telephone lines. The consumer experiences almost no degradation of service, and many feel that the cumulative effect of a large number of loads being shed at peak generating times can have a substantial effect on capacity growth.

The variety of short-run solutions that have been tried thus far have proved to be far from perfect, from the points of view of both the utility and its customers. Any one of these strategies brings with it multiple risks to the economy. First, virtually all strategies employed to date continue the policy of underpricing electricity to the ultimate consumer, and that underpricing encourages uneconomic use. In contrast, if customers were confronted with the real cost, they would probably choose to do something other than what they decide to do when they are charged the average price. This leads to a cycle in which capacity demands increase—more electricity is demanded than is economically efficient to produce—but rates lag.

Second, more capital investment may be required to meet future demands for electricity than would be the case if customers were provided with appropriate price signals. The resulting diversion of capital could, of course, mean that other, more efficient energy forms or nonenergy investments must pay higher costs or go unbuilt. Utilities do have a problem raissing capital, but this problem (discussed later in this statement) has little to do with theory and a great deal to do with political courage. Public utility commission policies, which have kept the rate of return to electric companies below that of other investments of comparable risk, have served to reduce the attractiveness of investing in utility equity or debt, thereby making undercapacity a potential problem in at least some parts of the country.

Third, because electric-utility rates are so clearly government controlled, rather than determined by marketplace forces either directly (probably an impossibility) or as a central goal of regulation (difficult but a possibility), there is a great temptation to begin using the electricity sector of the economy to accomplish a number of competing, explicitly social goals. Utility rate-setting procedures provide many opportunities to transfer enormous sums of money in complicated, often boring decisions that seem to deal only with technical rate changes. The money thus transferred serves to favor social goals that might not be favored if they had been stated explicitly. For example, electricity rate design can be used to lower the cost of electricity to users who consume only a small amount each month, a group that includes many of, though not all or only, the poor. Similarly, electricity rate design might be used to induce higher electricity consumption in order to shift away from imported oil. We believe that rates used for goals such as these cause more distortions than they cure and poorly serve their own policy objectives.

Some feel, for example, that lifeline rates can be useful within the context of a utility billing system, but we believe they are a poor substitute for real welfare payments. The tragic inability of this country to face welfare squarely as an obligation to be met rather than an issue to be dodged makes advocates for the poor look to lifeline rates as a way of quietly and automatically transferring some wealth from larger users of electricity to smaller users. In fact, the poor need money, not cheap electricity. But because getting them money to be spent for any purpose they might choose is so difficult, it is understandable that there are those who regard cheap electricity for the poor as an acceptable substitute.

Similarly, holding electricity prices to industry or to consumers below replacement costs, whether done explicity or through government-sponsored research and development or other subsidies, has been defended by some as a means of discouraging customers from substituting imported oil for electricity generated from coal or nuclear energy. As the past decade has demonstrated, electricity use responds to price, even in the short term, and higher prices mean that fewer kilowatt-hours are sold. Raising the overall level of rates will certainly push demand downward or reduce its continued upward movement. A number of observers believe that letting rates rise is unwise because they analyze the crux of the American energy problem as overdependence on imported oil, much of which comes from an unstable area of the world. Our analysis of the issue is rather different. As we see it, oil imports are an important national energy problem, but they are not the only one this country faces. The broader national energy issue is how we should use all energy forms more efficiently so that the economy will adjust into a configuration that better reflects changes in the real costs of each of its inputs. Unless those adjustments are made, the country will have completed the cycle of reindustrialization and recapitalization upon which it is now embarked on the basis of energy price projections that fail to recognize the real values of energy resources. That unsound rebuilding of our capital stock will, in our judgment, exact a toll on the economy that will ultimately lead to a diminution of our productive potential and our ability to compete in world markets.

A policy that uses lower prices of electricity as a means of keeping the country from moving away from domestically produced electricity (using coal and nuclear generation) and toward imported oil would provide no benchmark against which to determine when we had spent enough money on the electricity sector to justify the savings in imported oil. A more serious defect is that such a decision to hold down oil imports by means of increasing electrification almost certainly will mean that other, more cost-effective means of lowering oil imports will be ignored. Central among these options is more efficient use of energy. If oil imports are the problem, the solution is to restrict them directly by imposing import limitations (or indirectly by charging an import tax). To attempt to favor one particular domestic energy form as a means of reducing oil imports is a certain road back to severe economic distortions in the energy field.

LONG-TERM POLICY GOALS

We believe that long-term national policy should encourage a shifting of electricity pricing in the direction of a system based on replacement costs for all consumers. Such a shift should convey to all energy consumers a better sense of the resource cost to society of increased electricity consumption and should also provide to those utilities with a need to expand to meet growing demand the resources with which to do so. Furthermore, we believe that with enough imagination and determination, actions to accomplish this shift can be designed and introduced in a fashion that will not be seriously disruptive for energy consumers, distributors, or producers. But delay in getting started in the direction of rate reform serves no one's interests well.

Although we do not think the actual setting of rates should be federalized, we believe a number of steps could be taken to strengthen the ability of state regulators to reform rates in the directions we suggest. The directives contained in PURPA that utilities analyze their costs on a basis designed to reveal their replacement costs throughout the year served as a valuable stimulus for data gathering that will make further reforms easier. Any technical support or studies on load forecasting or demand-dampening strategies that could be provided more efficiently by academic institutions, foundations, the federal government, or other organizations to aid the more than fifty state commissions deserves support. The federal government could also provide support for states willing to experiment with the regulatory alternatives necessary to allow utility companies a more realistic rate of return and to create a methodology for taxing away any revenue above that level. These changes will not be easy to design or implement, but they are within reach. If we are to make progress in the energy field, it is very important that we begin to move in the direction of institutional reform.

Another long-term goal is to move toward deregulation of as much of the electric-utility industry as is feasible. There probably is little realistic possibility that the local distribution of electricity, at least to residences, can ever operate efficiently except as a monopoly (although there are some who feel that the ultimate promise of distributed fuel cells or photovoltaic cells may lead to independence from a grid). But it no longer seems as obvious as it once did that the generation of electricity must be a monopoly. Similarly, at least some bulk transmission of electricity may not need to be a monopoly. To a degree at least, PURPA requires electric utilities to buy electric power from cogenerators and small power producers and to pay these producers an amount equal to the costs of generating the power that the utility can avoid by not producing the power itself. Cogenerators of small quantities of electricity are given tax credits and are exempted from a variety of regulatory burdens.

There are many who believe that time-of-day pricing will encourage users able to do so to generate their own power during peak periods. Industrial consumers that have ready use for heat produced by their own processes may find it economical to make electricity as well as heat, relying on the utility company to distribute and resell the electricity. Thus, the structure of the industry as a whole may gradually change from one under the exclusive control of a single entity to a more mixed arrangement. This movement would be healthy, in our view, because power would then be produced and sold by those best able to make it cheaply. Resources would also be used more efficiently, with fuels formerly used only as heat sources being used to generate electricity as well. Of course, any change that is made must value appropriately the very high reliability of the interconnected electric system that now exists.

We believe that dispersing generating capacity, with the local utility company able to buy power from a variety of sources on the basis of its price and reliability, would give the industry a great deal more flexibility. It would also provide one way of putting an important component of costs beyond the reach of price regulation. The problems associated with a move away from utility ownership and operation of generating plants are complicated, involving both technical and legal issues, such as the preference clause that gives municipally owned utilities first call on all federal power generation. The responsibility for dependable supplies of electricity would have to be assumed by someone. Likewise, financing mechanisms are not easy to envision at this point. Nor will the transition from our present regulated structure to a more market-oriented one be easy. Nonetheless, the rewards appear to be substantial. The trend toward cogeneration is growing slowly in this country. Federal and state policies should continue to encourage active competition; both customers and the country as a whole will benefit.

Another important aspect of the price reform problem lies in the fact that it is linked to the multiple problems of earnings levels, "fairness," inflation, price discrimination, and the appropriate reliability and quality of service. These considerations and others bedevil regulatory commissions from all sides. Commissioners honestly trying to balance competing forces are understandably perplexed in their search for the "right" solution. Unfortunately, legal precedents give them little help. The conventional original-cost rate-making procedure, followed in most jurisdictions, is based on a 1923 opinion of Justice Louis Brandeis, who was writing during a period of price stability, an era that no longer exists. The use of this system accounts for much of the failure of most electric rates to come even close to reflecting replacement costs. It also accounts for the fact that, in many cases, today's rates are inadequate to maintain the financial stability of important parts of the electric power industry.

Although general inflation, the capital intensity of the industry, and soaring construction costs are important issues for the electric utility industry, one that is more within the reach of reform is the inability of the regulatory commissions to rethink their procedures and decision-making rules. If regulatory commissions are to deal more successfully with the problems of the industry they control, the issues of the adequacy of public utility rates of return and cash flow must be confronted directly and overcome by political courage. It is no long-term kindness to electric utility customers to set rates artificially low now if the mid- to long-term effect of that policy is to cause serious capacity shortages in the future.

CONCLUSIONS

Electricity is an important part of virtually any energy future one can imagine. Therefore, it is vital that the industry be healthy enough to support whatever load growth may occur. But it is equally important that whatever growth occurs be economically warranted in the sense that users would prefer to have additional electricity more than they would prefer alternative uses of their resources. Any overcapacity induced by incorrect price signals acts as a drag on the economy as a whole, foreclosing other, more productive investments.

We believe that the most important goal for the pricing of electricity is the same goal that ought to be pursued for oil, natural gas, and conservation: Prices to users should represent the marginal cost of the energy used and should confer the marginal benefits for energy not consumed. In the case of electricity, this will require a system that differs in many respects from the traditional, regulated system now in place.

Many have suggested that because marginal cost-based electricity pricing will, under present circumstances, cause rates to users to rise, any pricing strategy that raises rates is therefore desirable. A number of strategies that would also cause rates to rise—for example, inclusion of construction work in progress, increases in various reserve accounts, future test years, and faster recovery of capital expenses—have been urged in order to move the country toward the same rate levels that a marginal cost-based pricing policy would attain. Yet, the principles espoused in this chapter neither favor nor oppose higher electricity rates per se, even though their application under present circumstances would move rates upward. We do not find much to support in regulatory strategies that raise the rates without attempting to apply the underlying economic principles that ought to be followed. Instead, we urge that, to the greatest extent possible, electricity be accorded the treatment that is given to oil and that we believe should be given to natural gas; the rates charged will follow from application of that principle. Only then will the market serve its powerful function of sorting out what consumers really want and how much they are willing to pay for it.

CHAPTER FIVE

NATURAL GAS PRICING POLICY

It would be hard to find another market that has been as thoroughly distorted by regulation as the natural gas market in the United States. Purchasing arrangements, corporate structure, contract terms, and prices paid all reflect the pervasive and distorting effects of decades of regulation. Some of the laws enacted as recently as four years ago have made market conditions worse rather than better, demonstrating to most observers that even a complex, finely tuned regulatory scheme is no match for the unpredictability of the market or the ingenuity of the entrepreneurial spirit.

There is a great logical appeal to a return to basic principles in natural gas policy, a temptation that becomes increasingly attractive as we consider the complexities of the market as it now exists. Basic principles suggest that large industrial users should be able to bargain with gas producers (either directly or through brokers) for the purchase of gas under whatever terms of delivery and price they could negotiate, then arrange to move the gas from where it is produced to where they will use it either by a dedicated pipeline or through a common-carrier pipeline. Smaller users of gas would purchase gas from a local distributing company, a regulated monopoly that would act, in effect, as their agent for purchasing gas in volume from the producers in competition with others who want it. Such a market would more closely link producers and users through the price system. Producers and users alike would be free to buy or sell excess supply on the open market, hedge against the future as they thought appropriate, and otherwise act in accordance with their perceptions of their economic self-interest. Even in the case of pipelines, with their high fixed-capital costs, owners could take whatever protective arrangements they thought necessary to secure financing.

How far from this simple model we have strayed. Instead of a system that brings producers and ultimate users close together, we have developed a system that insulates producers and users from receiving supply or demand information from one another. To the extent that any price information passes through the institutional barriers that regulation has built, it is further obscured by a web of price controls that even many experts find baffling.

A short review of the current situation of each of the four parties to the market (producers, pipelines, distribution utilities, and ultimate consumers), together with a history of how the situation decayed to this point, will provide a background to understanding a set of policies that may help move the market back toward a more rational arrangement.

WHO IS AFFECTED BY NATURAL GAS PRICING?

PRODUCERS

Until 1954, the price at which natural gas was sold was largely determined by prices negotiated between producers and pipelines willing to transport gas. Natural gas, frequently found during the search for oil, was an inexpensive source of energy; and as it became more widely available, it gained a healthy share of the market. In 1954, the Supreme Court ruled that the Federal Power Commission (now the Federal Energy Regulatory Commission, FERC) had legal power under the Natural Gas Act of 1938 to set the price paid by pipeline companies engaged in *interstate* sales for resale of gas; *intrastate* sales of gas were not covered by the act and therefore remained unregulated by the commission. As a result of the Supreme Court ruling, two separate markets for natural gas developed during the twenty-five years between 1954 and the passage of the Natural Gas Policy Act (NGPA) of 1978. Interstate sales during that period were set at prices that represented what the commission believed was a fair return on the cost of producing the gas, regardless of the prices of alternative energy sources. Under a legal standard that required prudent behavior, the commission also had the power to regulate terms and conditions of gas sales, but there was relatively less interest in regulating those hard-to-quantify items. Thus, in interstate transactions, negotiations between pipelines and producers focused on nonprice terms such as delivery volumes, what would happen if the purchaser did not want all the gas covered by the contract, and how frequently and at what level the contract might be renegotiated. In contrast, intrastate market sales varied enormously both in price and in other terms and conditions, reacting to the operation of a free market in which buyers and sellers could negotiate.

Between 1954 and 1978, the relationship between the prices of interstate and intrastate gas sales varied. But during the 1970s, probably as a result of the pressures of the Clean Air Act, higher oil prices, and the increased manufacture of petrochemicals and fertilizers based on gas as a feedstock, the price achieved in intrastate sales soared above the regulated interstate price (see Figure 5). Not surprisingly, shortages developed in the interstate market as producers favored wells that could sell gas within the boundaries of the producing state and therefore be free from federal price controls. Bans on new gas hookups were imposed in some jurisdictions. The cold winter of 1976–1977 brought the situation to a head politically when low-priority factories closed because state utility commissions, following curtailment plans, cut off some industrial customers in order to ensure continued gas service to homes.

This two-tier market and the dislocations that it helped induce led Congress to pass the NGPA, which fundamentally altered the regulation of all natural gas. The principal objective of the NGPA, and the part that caused the most heated debate, was gradual decontrol of the price of *new* (post-1977) natural gas at the wellhead over a seven-year period. Formerly unregulated intrastate gas was subject to federal control for the first time; *old* interstate gas and certain intrastate gas were permitted some price increases but were not allowed to move toward decontrol. The legislation was based on the assumption that natural gas is a scarce and uniquely valuable resource; the objectives of the act were to ensure that the price of natural gas should track that of world oil and that discoveries of certain types of hard-to-find natural gas should be rewarded with higher prices. These objectives,

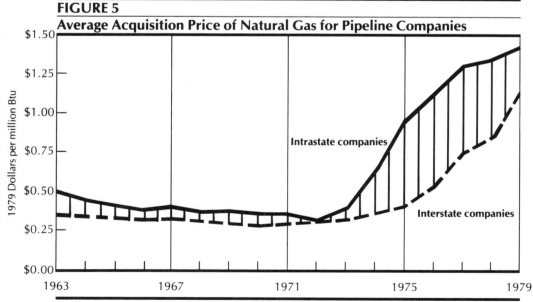

FIGURE 5

Average Acquisition Price of Natural Gas for Pipeline Companies

SOURCE: U.S. Department of Energy, Energy Information Administration, *Monthly Energy Review* (Washington, D.C.: U.S. Government Printing Office, January, 1982), p. ii.

which a smoothly functioning market performs automatically, were over-laid with a social policy objective of protecting residential and commercial consumers from most of the initial burden of price increases by shifting the higher costs to industrial users first.

The NGPA created a large number of categories of natural gas and set a different price ceiling for each. Generally speaking, old gas (gas flowing prior to 1977) is to be priced at the amount for which it sold in November 1978; only adjustments for inflation are allowed. The price of new gas (gas produced after early 1977) is allowed to rise at the inflation rate plus a series of increases designed to track the assumed world oil price. Congress be-lieved that oil would sell for $15 a barrel in 1985, so the scheduled in-creases let new gas prices rise to that level (on a price-per-British thermal unit basis) by 1985, at which time price controls on new gas will be re-moved. *Unconventional* gas (gas produced from certain specified depths or types of formations that have only recently become commercially intrigu-ing) is subject to no price controls after November 1979.

The overall effect of these price regulations is that in 1985, according to FERC estimates, about half the gas flowing in pipelines will be old gas selling at the inflation-adjusted equivalent of 1978 prices; the remainder will be gas that has risen only to the Btu-equivalent price of $15-a-barrel oil. In 1985, however, the new gas will be completely decontrolled and will probably rise sharply in price (the so-called spike), but most of the old gas will remain under price regulation.

Consequently, at least until supplies of old gas are depleted, there will be a great advantage to those users who have access to large amounts of the old, price-regulated gas either to use it exclusively or to mix it with small quantities of higher-priced gas to produce a lower average cost (a pricing strategy referred to as *rolling in*). Pipelines that have access to large quanti-ties of the old price-controlled gas will be able to bid up the price of uncon-trolled gas higher than their competitors can afford to pay because their av-erage selling price of the old-new gas mixture will still be no higher than the going price of gas. Therefore, even after decontrol occurs in 1985, the price averaging possible with old gas will provide a cushion whose effect will be to permit decontrolled gas to sell for a premium related, not to its value, but to how much cheaper old gas the seller is able to mix with it.

Adding to the distorting effect of this intricately controlled NGPA non-market is the fact that old intrastate gas (which became subject to regulation for the first time under the act) is treated differently from old interstate gas, leading to a situation in which the interstate market is likely to have a larger cushion of price-controlled gas than the intrastate market. Thus, different producers and transmission companies will receive rewards that are based on government regulation rather than on their abilities as businessmen.

The second objective of the NGPA, to provide protection for residential and commercial users, is to be accomplished through a system called *incremental pricing*. As natural gas prices are permitted to rise through the operation of the NGPA, the law requires that all prices above a reference price be allocated by distributing companies to certain industrial users up to a trigger point before any of the higher prices are passed on to residential consumers. The trigger point is that price at which industrial customers are likely to switch to alternative fuels rather than pay the price of natural gas. Unfortunately, this system has had disruptive side effects. The price of alternative fuels that serves as the basis for determining the trigger price is not easy to calculate, so that fixing the trigger point is subject to error in both directions. There is an even more subtle effect as well: The very existence of legislation that contemplates offering more protection to residential users provides a message to industry to switch from natural gas to alternative fuels if the prices are reasonably close. Fuel switching by large industrial consumers can have an adverse effect on the prices paid by small users who are less able to switch to other sources. As the fixed costs of distribution and operation are spread across fewer units of gas sold, each unit becomes more expensive, causing further desertions from the system and leaving residential and commercial customers to bear higher costs based on lower demands. Thus, although the system of incremental pricing may have short-term advantages to the residential consumer, it could cause unexpectedly large price rises.

At the time the NGPA was passed, many producers welcomed it as promising an end to regulation. But as the scheduled price increases have fallen increasingly behind the Btu-equivalent price for world oil, producers have become correspondingly unhappy with the rewards offered them for new production. There is considerable debate about whether supplies would increase if new gas prices were decontrolled now. A study commissioned by the Natural Gas Supply Association, a group of gas producers, concluded that domestic non-Alaskan gas production could increase to the point that additions to reserves would equal consumption but that "such a turnaround has not yet occurred, nor is it likely to occur under the NGPA. If such a turnaround is to occur, it will require substantial improvement in economic incentives which accelerate drilling activity significantly beyond the levels which are likely under the NGPA."[1]

---

[1]/Edward W. Erickson, *Natural Gas and the Natural Gas Policy Act: A Pragmatic Analysis* (Washington, D.C.: Natural Gas Supply Association, January, 1981), p.III.

In contrast, other observers believe that drilling activity is taking place as quickly as equipment and manpower permit. According to this view, the opportunities for increasing the domestic supply of natural gas are extremely limited, and therefore permitting higher prices to be charged for old gas simply provides economic rents (windfalls) without any prospect of calling forth significantly more supply.

Because the debate depends on physical facts that can be guessed at but not known and on economic principles not universally trusted in the political process, policy decisions must be based partly on educated guesses concerning availability of more gas and the effect of incentives on producers and partly on the decision maker's own philosophy concerning the relative merits of free-market forces and regulatory or planning actions.

Although the uncertainties are significant, we are persuaded that the potential sources of methane in this country are quite large and that there is every reason to believe that higher prices could stimulate production of large amounts of conventional and unconventional gas. On the philosophical side of the debate, we believe that in the face of uncertainty, decisions made according to a priori assumptions about future gas reserves are a poor basis for imposing price regulations.

To underscore this point, let us consider the two extreme possibilities. On the one hand, suppose that potential supplies are large. If these new supplies had been discovered and used if prices were allowed to rise but controls prevented the new production, the cost to the nation would be enormous in terms of lost opportunities to avoid additional oil imports, high-priced gas imports, and increased dependence on electricity to perform tasks for which gas is suited. On the other hand, suppose that no additional gas remains to be discovered at any higher price. Even so, deregulation has the advantage of stretching gas supplies so that consumers will have earlier, more accurate signals to make the investments necessary to find alternative fuels or use less gas, devoting the limited supplies to the most highly valued uses. Thus, even though we believe that more natural gas could be produced at higher prices, this belief is not central to our urging that producer prices should be decontrolled. Even if our optimism proves incorrect, decontrol will encourage conservation and fuel switching away from gas in a smoother fashion than would be possible if price controls persist.

PIPELINES

Unlike railroads or trucking companies, gas pipelines do not operate simply as transportation companies. Instead, interstate transportation of natural gas is accomplished by companies that buy the gas from producers at the field and then transport it in bulk, splitting off on spur pipelines at various points to deliver gas to other companies (typically local distributors). When the pipeline companies first began to market gas twenty or thirty years ago, they offered long-term contracts to the distributing companies, often at low rates that reflected both the relative value of natural gas at the time and the desire of the transmission companies to ensure a stream of revenue that would justify the borrowing necessary to construct the pipeline facility.

Since 1954, the price that pipeline companies pay for gas at the wellhead has been controlled by federal regulation. As a result, the price component of the bundle of rights and obligations that goes into making up any sale could no longer be negotiated. But a variety of other important terms remained: How much gas had to be taken and for how long? What future events would trigger a change in the initial price, and how large a change could be made? What events would permit one party or the other to terminate the contract or renegotiate its terms? By law, these terms can be controlled by FERC, but in fact, they are less subject to regulatory oversight. The producer and the pipeline had somewhat different interests, although the interests of the pipeline were not those of the ordinary consumer; many of the items that might be negotiated had simply to be accepted by the pipeline's own customer, the distributing company. Generally, producers wanted to sell gas from newly discovered fields fast in order to start receiving a return on their investments. Therefore, they tried to get the pipeline to agree to take a large proportion (perhaps all) of the gas produced at the field at the agreed-upon price. The pipeline, if it could not sell all the gas it was obligated to purchase, might incur penalties (which could generally be passed on to its customers). The pipeline may have been able to negotiate favorable price-escalation clauses with intrastate producers, taking advantage of the pressure to produce felt by the well owner as an inducement to hold down future price increases. But in the interstate market, where escalations were governed by regulation, that leverage was denied to the pipeline. In short, when the price to be paid was, by law, placed beyond negotiation, the producer had considerable power to exact other terms from the pipeline that could add to the total cost of the gas to the ultimate user.

The pipeline system is currently operating below capacity, although the availability of excess transmission capacity is subject to a great deal of regional and seasonal variation. Moreover, transmission costs, both fixed and variable, have been falling steadily over the past decade as a percentage of the total cost of the delivered product, even though virtually all cash expenses of operation have risen during the period. The pipeline companies have a great interest in keeping a high volume of gas flowing in the pipeline so that the fixed costs are spread over as much gas as possible. It is not surprising, then, that the pipeline companies fear that wellhead price deregulation, to the extent that it encourages either conservation or switching to alternative fuels, will cause the unit price of natural gas to rise to cover the fixed costs and thus drive customers to further switching. However, because of the way in which pipelines are regulated at the federal level, the pipeline company itself would not suffer any immediate adverse economic effects from higher gas prices because it is permitted, under purchased-gas adjustment clauses, to pass on to its utility customers the full cost of either expensive gas or any penalty it incurs under a take-or-pay provision.

Pipeline companies provide a variety of useful services to the gas-consuming community. Because they actually own most of the natural gas they sell, they are the major active participants in planning for the continuity of gas supplies. It is in their interest to make sure that their facilities are used as fully as possible and that their customers, the distributing companies, receive dependable service. In the absence of other institutions to plan for continuity and thus stabilize the market, the pipelines have served that function in the interstate market. At the moment, no other institution is ready to undertake it, and many fear that leaving this role solely to millions of market decisions would be extremely disruptive. The institutional structure of the natural gas supply system and the rates that pipelines charge are both topics worthy of study, but they lie beyond the scope of this statement.

DISTRIBUTION COMPANIES

The approximately 1,476 distribution companies in this country are equally divided between privately and publicly owned entities, although their sales volumes are not equally divided. Distributing companies, some of which provide both gas and electricity within their service territories, buy most of their gas from pipelines and sell it through a system of buried pipes to ultimate users within given geographic areas. Some distributing companies provide other services, such as storage of relatively small volumes of gas to dampen small swings in demand, manufacture of so-called peaking gas for use in extended cold periods, and other small importation or manufacturing services. Because these high-cost activities to lower peak demands made on the pipeline supply of gas enable the utility to avoid a higher demand charge, they may save money overall for the utility's customers. Pipelines typically impose a demand charge based on the highest volume purchased by the distributing company during the year as a way of passing on a portion of their capacity costs to utilities that have great variations in demand. Gas utilities typically do not attempt to impose that demand charge directly on those customers who cause the surges in demand. In the absence of such targeted rate setting, even a relatively expensive peaking facility may be less expensive overall for the distributing company and its customers than paying a high demand charge for a year.

With two exceptions, the details of rate setting at the retail level are not pertinent to this statement. First, as in the case of the pipeline companies, the percentage of the final price represented by purchased gas has risen over the past decade, and most estimates suggest that it will continue to rise. According to a Department of Energy study,[2] the fixed costs of both pipeline and distribution companies represented about 80 percent of the final sale price of gas in 1955; in 1978, these charges were only 60 percent of the price; by 1990, it is estimated that they will represent only about 30 percent. But the distribution companies share the pipeline companies' concerns about the decline in volume that would cause the same fixed costs to be spread over fewer units of higher-priced gas. Holding on to a customer base means revenue stability, an extremely important goal for companies that must plan over a multiyear period. Therefore, they have paid increasing attention to the competition for market share in which they find themselves engaged and have become active in discussions of state and federal policies that may endanger the volume of their sales.

Second, retail rates represent an average of various prices of gas that the pipeline may purchase and sell to the utility at a homogenized, or rolled-in, price. The rolled-in price of high-cost gas is passed on directly to the consumer through the purchased-gas adjustment that appears on the utility bill, raising (or lowering) the price per unit of gas sold.

The net effect of these two features of retail rate setting is that the utility is less price-sensitive than it might otherwise be when it purchases gas from the pipeline because it can pass through its major cost (gas it purchases or makes) without approval of the state regulatory commission. At the same time, the ultimate consumer is not at all aware of the costs imposed on the system by increased demand at certain times (e.g., a cold winter day) or by suddenly using volumes of gas that vary from the average by a great deal. This further insulates the consumer from real and rising costs and denies all customers the information they could use to shift or reduce gas consumption and lower their own total costs.

The distributing companies see an enormous potential for new gas production, and they speak of natural gas as being both a bridge fuel to the

---

[2]/*Public Utility Regulatory Policies Act of 1978: Natural Gas Rate Design Study*, DOE/RG-0035/1 (Washington, D.C.: Economic Regulatory Administration, U.S. Department of Energy, May 1980).

future and a renewable resource. But some of them fear that short-term regulatory actions that hold down use of natural gas by some consumers or actions that serve to pass on the costs of high-priced gas production immediately to the consumers will start a downward spiraling of demand that will feed upon itself until gas is no longer able to compete in the energy market.

ULTIMATE CONSUMERS

About 27 percent of all the energy consumed in the United States comes from gas. Natural gas is the primary energy source for more than half of all residential and commercial establishments, serving some 43 million households. Nearly 40 percent of all energy consumed in the industrial and agricultural sectors is natural gas. Gas is an important feedstock for the production of petrochemicals and agricultural fertilizers and chemicals. Electric utilities, particularly in the Southwest, also consume considerable quantities of gas. The industrial sector is the largest consumer of natural gas (see Figure 6).

Traditionally, natural gas was sold to any consumer for any purpose. But during the 1970s, concern over natural gas supplies grew, and the shortage in the interstate market created by price regulations necessitated a variety of restrictions on certain uses. Some states forbade the use of decora-

FIGURE 6
Consumption of Natural Gas by End-Use Sector, 1981 (Quadrillion Btu)

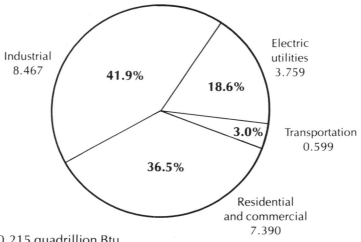

Industrial
8.467

41.9%

Electric
utilities
3.759

18.6%

3.0%

Transportation
0.599

36.5%

Residential
and commercial
7.390

U.S. total: 20.215 quadrillion Btu

SOURCE: U.S Department of Energy, Energy Information Administration, *Monthly Energy Review* (Washington, D.C.: U.S. Government Printing Office, February, 1982), pp. 22, 24, 25.

tive gas lanterns; that was later made federal policy as well. Curtailment plans established complicated priority systems under which electric utilities and large industrial users were severely restricted in their use of gas in periods of shortage. Some states prohibited hookups of new customers. A number of industrial customers that had enjoyed slightly lower rates in exchange for agreeing to have their service subject to interruption found that the interruption clauses were invoked; as a result, workers were laid off. No limits were imposed, however, on demand from residential consumers.

Federal law added significantly to these end-use controls in the Powerplant and Industrial Fuel Use Act of 1978 (PIFUA). The act's stated purposes include conservation of natural gas "for uses other than electric utility or other industrial or commercial generation of steam or electricity" and encouragement of "greater use of coal . . . in lieu of natural gas and petroleum as a primary energy source."[3] Basically, and subject to a large number of exceptions and qualifications, PIFUA makes it unlawful to build a new electric plant or industrial boiler that uses natural gas as a fuel. Natural gas, according to the philosophy implicit in the act, is too valuable to be used where coal might be substituted, and the limited supply of natural gas must be saved for the use of residential and commercial customers, who have the hardest time switching to alternative fuels.

At the same time that PIFUA was enacted, the NGPA became law as part of the National Energy Act of 1978. It created a policy of incremental pricing that placed the burden of higher gas prices on industrial users first.

The provisions of PIFUA have not been strictly interpreted, nor have they been implemented as quickly as Congress might have expected in 1978. Nevertheless, PIFUA and the NGPA together distort the industrial market for natural gas by offering at least the possibility that both pipelines and distributing companies will be deprived of their high-volume customers, who serve to balance the residential users' seasonal consumption.

At the same time, even though the price of gas to end users is still below its replacement cost, the price increases have been large enough to induce some dramatic improvements in end-use equipment. Notable advances have occurred in the gas appliance market. For example, pulse-combustion furnaces that offer seasonal efficiencies of around 92 percent, compared with conventional furnaces, which have efficiencies around 60 percent, will be on the market this year. The fuel cell stands nearly ready to provide both electricity and heat to apartment and office buildings. Industrial heat

---

[3]/Powerplant and Industrial Fuel Use Act of 1978, PL 95-620, enacted October 14, 1978; signed by President Carter November 9, 1978; Section 102 (b) (2) and 102 (b) (3).

pumps, waste-heat recovery systems, and advanced industrial burners promise industrial improvements in end-use efficiency. Thus, there is considerable evidence that if the price of natural gas rises, consumers of all kinds will find ways to cut their use of the fuel.

THE NATURAL GAS DILEMMA

Moving to a market system for pricing natural gas will be complicated, and the choices involved are by no means simple or straightforward. The natural gas policy dilemma may be summed up in a series of statements and counterstatements:

- A policy of complete decontrol would result in price increases to gas users. If higher prices cause massive fuel switching away from gas by industrial consumers, those who were not able to switch (principally the residential users) would have to pay increasing rates as they absorbed more of the fixed cost per unit consumed.
  On the other hand, economic policy suggests that consumers should pay the full marginal cost of the energy they use so that they will be motivated to make rational investment and purchase decisions. Unless they pay those prices, they will consume too much gas, make capital-purchase decisions that lock the country into a high-cost energy future, and deplete gas supplies sooner than necessary.

- The windfalls (or economic rents) from decontrol of natural gas could be given to consumers or returned through the tax system.
  On the other hand, returning economic rents in any fashion that depends on administrative calculation of true or average costs is very likely to be erratic. Any such system would surely impose a significant regulatory burden and exact substantial efficiency losses.

- Price signals should pass information back and forth between buyers and sellers.
  But the structure of the industry and its regulation makes such exchanges difficult in the absence of institutional changes.

- Some of the future methods of producing more natural gas or transporting it appear so expensive that consumers would not want to buy the gas unless the price is rolled in with less expensive price-controlled gas.

On the other hand, rolled-in pricing permits the financing of projects whose economic ends might be met more cheaply through other means, particularly improved end-use efficiency.

- Efficiency in the use of all fuels is essential if this country is to prosper as a leading industrial nation in the next century.
  On the other hand, some fear that fuel switching may reduce the gas market to such an extent that a high-value, domestically produced fuel will not be fully developed.

- Gas price decontrol is an immensely complicated business because of the complexities resulting from very many federal, state, and local government regulations and the inflexibilities that arise from capital investments by both producers and consumers. An additional inflexiblity arises from long-term supply contracts. Many such contracts signed by industrial users carry price-escalator clauses that might produce an exaggerated price increase upon decontrol. The transitional adjustment to decontrolled gas prices could be painful in many cases.
  But the longer gas decontrol is delayed, the greater the overall loss to society through accumulated inefficiencies in energy production, distribution, and consumption.

A NATIONAL POLICY

On balance, we believe that the advantages of deregulation in promoting efficient use of natural gas and in encouraging production outweigh the possible drawbacks. To be sure, we recognize the magnitude of the costs involved in any transition to a deregulated environment and the need of consumers to have time to make defensive investments that will cushion them from the shock of significantly higher prices. We also feel that certain reforms in the regulation of utilities will add considerable efficiency to the natural gas markets.

PRICE REGULATION AT THE WELLHEAD

We believe that the current price regulations for gas under the NGPA are both unrealistic and unwise as a matter of national policy. We believe the economy will adjust to deregulation of natural gas prices, just as it adjusted to the deregulation of oil prices. Under the current structure and pace of price increases in world oil, there is bound to be a sharp price hike in 1985 in any case as new gas is decontrolled. We believe that delaying that increase makes little or no sense. We endorse decontrol covering all categories of gas.

UNCONVENTIONAL GAS. Under the NGPA, gas that comes from sources such as wells tapping formations lying 15,000 feet or more below the surface, Devonian shales, coal seams, and geopressured aquifers is completely free from price controls at the wellhead. We believe that this policy is correct. The risk and the expense of locating and developing these unconventional gas sources are so great that the sources would not be developed if price controls were extended to them. However, we also believe that no direct or indirect subsidy should be granted for exploration or development of these gas sources. Instead, they should come into production only at the rate that the market permits. Many of these sources now command high prices because the price of unconventional gas can be rolled in with that of price-controlled gas. Taken together, our recommendations would no doubt slow down development of these sources until gas commands a higher price in the market. But this sorting out of the order of investments in new production is one of the major benefits to be derived from decontrol.

NEW CONVENTIONAL GAS. New gas (i.e., gas produced from proven reserves after early 1977) now represents a relatively small percentage of the market; by 1985, however, it may be around half of the market, and by the 1990s, it will represent most of the flowing gas. Under the NGPA, the price of this gas will jump suddenly in 1985 from about $2.70 per million Btus (equivalent to oil at $15.70 a barrel) to whatever energy in the form and quantities that natural gas can deliver commands in the open market. (Many observers do not believe that natural gas will rise to the Btu-equivalent price of oil because of transmission and distribution costs and the lack of demand at such a price; others believe that it may exceed the oil-equivalent price because of its cleanliness and ease of consumption. Therefore, we make no estimates about the size of the price jump in 1985 for new conventional gas that is then decontrolled, although we are confident that there will be a jump and that it is likely to be substantial.)

Just as we do not believe that it is in the consumers' or the nation's long-term interest to pretend that oil is less valuable than the price it might command in a market, we do not believe that such underpricing makes sound natural gas policy. Therefore, the question is whether it is better to take gradual price increases, with a sharp rise when controls are lifted, or to take a price shock all at once. Some argue that getting inflation under control is so important that the nation should not risk the addition to the CPI that would result from decontrol of the price of new gas. We find this argument unconvincing. It is important to achieve the efficiencies that better price signals can bring earlier rather than later. Only by making investments in response to the correct signals can one of the underlying causes of inflation, misinvestment, begin to be corrected. Because the price shock is inevitable, we are persuaded that it is better to take it sooner rather than later. We do not believe that preserving the market share for natural gas is a sufficient reason to continue to use regulation to hold prices down.

OLD GAS. Under the NGPA, old interstate gas is regulated at its 1978 price plus adjustments for the rate of inflation. Supplies of this old gas will not be exhausted until the 1990s. Therefore, old gas, which will remain controlled until it is used up, provides a cushion against average price increases, but in doing so, facilitates still greater increases in the prices of uncontrolled gas. It is important to recognize that there are two reasons for the existence of the cushion. First, the law regulates the price at which the old gas may be sold; no sale is permitted at a higher price. Second, the price for a varying percentage of old gas is also set in the contracts between the producer and the pipeline. This feature of regulation makes determining the effect of lifting price controls from old gas especially difficult. Some contracts contain deregulation clauses that would permit the price to rise immediately; others contain most-favored-nation clauses that tie the price to the best price the pipeline offers from that field or in general; still others name a price; many are silent on the issue. Apparently, many contracts are so ambiguous that unless the parties can agree between themselves, the price level might be settled only after a lawsuit. FERC has attempted to estimate what would happen if old gas were decontrolled and has concluded that no rational basis for prediction exists.

Some distributing companies and pipelines fear that if old gas is decontrolled, producers will invoke their contracts to force the pipelines and distributors to pay above-market-clearing prices for their gas, even if it cannot be sold at that high price. These observers predict that there will be a major loss of customers as industries convert to oil, that distribution companies will be forced into bankruptcy as state utility commissions refuse to permit full costs to be passed on to gas users, and that imported oil will be substituted for major portions of domestic gas consumption that cannot, because of the operations of contract provisions, be sold in the market.

We recognize that the old gas now flowing under existing contracts containing take-or-pay provisions, most-favored-nation clauses, and deregulation-escalation sections poses some of the most difficult transitional problems in the natural gas area. In a worst-case scenario, deregulation could trigger clauses in contracts that permit all gas sold from a field to be priced at the very highest rates charged for any gas from the field, even if that very expensive gas was put into production only because it had received special incentives under the NGPA. Under existing practice, those prices, even if above market-clearing levels, would be supported by regulatory pass-throughs.

On the other hand, we are mindful that in trying to solve this important transitional problem, there is a strong possibility that the country could find itself backed into either continued price regulation or legislative solutions that would be of enormous complexity and uncertain results. We are not entirely convinced that contract terms will, in fact, be used by gas producers to destroy the only buyers they have for their product. We urge Congress to resist developing detailed and intricate solutions to the problem of existing contracts. Instead, we believe that more thoughtful legislation should be limited to devising a system that encourages private renegotiation. Because we recognize that continued regulation of the pipelines and the distribution companies can serve to reduce the incentives for vigorous negotiations among the parties, there may also be a need for inducements for all parties to renegotiate in a cost-minimizing manner.

Regulation that is as longstanding as that in the gas field will not be easy to unravel. In the future, the way that government regulates pipelines and distribution companies may require substantial reform before all the benefits of wellhead deregulation can be achieved. But we believe that Congress should not become involved in detailed rewriting of private contracts as part of wellhead deregulation. We also believe that the problem of existing contracts is so intractable that it should not be used as a reason for moving forward on deregulation.

From a policy viewpoint, there is certainly nothing to recommend in treating some old gas as if it is worth less than chemically identical new gas simply because of the time at which production at a field began. The reasons that appealed to us for earlier decontrol for new gas are also attractive in the case of old gas, although we recognize that the process will not be as orderly or the price hike as sharp because of the uncertain number of contracts that must be individually renegotiated or reinterpreted. The process would reduce the amount of cheap gas that could be rolled in to cushion the transition and begin the process of bringing the full replacement cost of gas to bear on the consumer, which we believe should happen sooner rather than later.

The issue of a windfall profits tax must also be considered. The political price of decontrol of crude oil was that already-producing domestic oil wells had to pay an excise tax. The windfall profits tax apparently made good political sense and was the key to freeing the price of oil. Unfortunately, the tax as enacted was an excise tax, not a tax on economic rents.

Moreover, it applied to new oil as well as old and thus did not carry out the policy that its name suggests. It is at least likely that a windfall tax on old gas might be the price required for natural gas decontrol. We believe that raising the price of old gas as fast as contracts will permit is so important that if a tax is the price of achieving that goal, it is a price worth paying. However, we do not favor such a tax, particularly a tax that is as ill targeted as the one on oil.

Several important principles should be followed if a windfall profits tax is to be part of the decontrol package. The tax should apply only to old gas, which should be defined in a way that does not discourage new drilling in existing fields. The tax should be phased out with the depletion of supplies of old gas. And the proceeds of the tax should go into the general revenue fund instead of dedicated trust funds. Having said this, we repeat that although we have reservations about a windfall profits tax, we would accept it if it is the necessary price of decontrol, and we urge that it be truly related to capturing economic rents, not simply generating revenue.

One final issue, that of rolled-in prices, needs to be discussed. Price averaging occurs in at least two situations. The first is the case in which a pipeline purchases some conventional gas that is expensive and some that is inexpensive. Theoretically, all the gas should be sold at the price of the most expensive; but in fact, some measure of price averaging as a result of private-party negotiations is probably inevitable because gas is sold under a patchwork of older and many newer contracts from older and newer fields and by smaller efficient producers or larger inefficient ones. This price averaging will occur, not through regulation, but through negotiations between buyers and sellers. The rolling in of prices becomes unwise policy when it is used under the force of law or regulation to lower the apparent price of an uneconomic production or transmission facility. For example, rolling in the price of a high-cost synthetic gas plant or an expensive pipeline conceals the true cost of the gas and its transmission from both the consumers and those who might produce gas from other, less costly sources. We believe that the use of rolled-in pricing to support such large projects denies consumers the choice of using less gas or the same amount of gas more efficiently.

Rolling in prices can also have an unhealthy effect on producers by insufficiently rewarding production of gas that would have to sell for only a little bit more than the average price and favoring production of more expensive gas whose true cost is hidden by price averaging. In this way, supplies may become available, not in order of economic attractiveness, but in some fashion that is less attractive for future users. One of the greatest advantages of deregulation may well be that it tends to remove the opportunity to subsidize expensive gas supplements by rolling such gas in with artificially cheap, regulated gas.

With deregulation, the rolling-in problem will become less severe. The potential for abuse will be reduced but not entirely eradicated because even under deregulation, there will be some old gas sold at a low price under old, fixed-price contracts.

DISTRIBUTION COMPANIES

The design of retail natural gas rates is controversial. Under PURPA, the Department of Energy was charged with studying the rate structure for sales of natural gas. It concluded[4] that distribution utilities should move away from traditional rates based on accounting costs (the original cost of the plant and distribution system as depreciated) and toward rates that represent the economic value of natural gas to the consumers. Under the rate structures proposed in the study, customers would pay for gas at rates equal to the cost of supplying additional gas. As we explained more fully in Chapter 4, in cases in which this practice causes an overcollection of revenue for the distribution utility, lower charges for the earlier units of gas consumed could be used to absorb the excess revenue. Rate structure could be designed to produce utility bills for the average customer in each rate category that would be the same as those rendered before the change. For the user who consumes more than his or her class average, bills could be significantly larger; for small users, they could be significantly smaller. (The study also made it clear that fundamental changes in the direction of replacement-cost pricing could not be accomplished at the distribution-company level unless pipeline rates were also changed.)

The Department of Energy study was not greeted with universal favor by the natural gas transmission and distribution industry. Nonetheless, the general approach that it recommends is one with which we feel comfortable. There is no other rate structure to the ultimate consumer that accurately conveys the economic value of saving gas at the margin (which is where decisions on how much to consume are made).

Natural gas rates at the retail level are not set by the federal government; they are regulated by state and local utility commissions or by boards in the case of publicly owned utilities. Therefore, unless the federal government extends its control over local rates by legislation (an action that would probably be upheld as constitutional, although it would also probably lead to an administrative snarl reminiscent of gasoline price controls and alloca-

---

[4]/Public Utility Regulatory Policies Act of 1978: Natural Gas Rate Design Study, DOE/RG-0035/1.

tion), the pressures for such retail rate reforms will have to remain at the state level. At that level of decision making, changes may be uneven and slow. We urge that commissions and distribution companies move toward rates that reflect the economic value of gas, at least in the final units of consumption.

CONSUMERS

The short-run reforms that are necessary at the users' end are less controversial. There is no reason to set legislative limits on the uses to which natural gas can be put; the PIFUA provisions to that effect should be eliminated. Gas should be sold to those users who are willing to pay its price for the uses that they themselves choose. An energy policy that dictates who may use what fuel type is particularly inappropriate for natural gas because of the great uncertainties about reserves, their locations, and their recoverability.

MITIGATING THE EFFECTS OF IMMEDIATE DEREGULATION

To evaluate the effects of immediate decontrol of all natural gas requires at least some estimation of whether the shock of suddenly higher natural gas prices would have a devastating impact on the economy as a whole. If many people believed that it would, the policy would have little or no chance of being adopted because such devastation would be politically unacceptable, not to say unwise.

Unfortunately, it is not easy to answer the question of cost very accurately. In this country, we have virtually no experience with real national markets for natural gas. Although the intrastate markets that existed before the NGPA eliminated them gave some clue to the market value of gas, that information is suspect because of the artificial nature of a market constrained by state boundaries; moreover, it is now out of date. Similarly, the sales of gas from Mexico and Canada that are now occurring do not represent a market price because the buyers can roll in those high prices with the price of domestic controlled gas. Moreover, if old gas is decontrolled, an uncertain amount of that gas will be priced, not by law or by operation of today's market conditions, but by long-term contracts either as entered into as long as two decades ago or as modified under strong pressures from both sides.

The fundamental reason for the uncertainty concerning the total effect of any strategy of natural gas deregulation is that no one knows how deregulation will affect the quantities demanded. Because price and demand are as inextricably linked as price and supply, sharp increases would surely drive demand down. There might be some lag while gas users made adjustments in reaction to the higher prices, but consumer response in other sectors, such as automobiles, has demonstrated that demand can drop much faster than experts might predict. In fact, the period of readjustment is far more likely to be measured in months than in years. As demand dropped, prices would fall until a new equilibrium was reached at which producers and consumers struck the most satisfactory deal possible. (A contrary, demand-increasing trend is at least possible. Many believe that there is a great deal of pent-up demand for gas among those users who are now prevented from having it because of the operation of federal or state law and regulation or who are discouraged from hooking onto the system because of the supply uncertainties that priority categories impose on large users. Thus, the price-induced demand drop might be offset, at least in part, by new demand.) At that point, the market would clear, and a price would emerge. But there is no way to guess in advance what that price might be.

In spite of all these difficulties and uncertainties, we can acknowledge that *at least in the short run,* the extra price that consumers of gas will pay to producers of gas might be extremely high. The Interstate Natural Gas Association, a group representing interstate pipelines, favors lifting restrictions on use of natural gas but opposes any changes in the NGPA pricing provisions. The association has estimated that deregulating only new gas on January 1, 1982 (instead of on January 1, 1985, as the present act requires) would have added $20.7 billion to gas prices in 1982, $27.7 billion in 1983, and $34.6 billion in 1984.[5] Deregulating both old and new natural gas on January 1, 1982, would have added $72.8 billion to gas prices in 1982, $76.7 billion in 1983, and $81.8 billion in 1984. (The highest estimate for 1984 is approximately $30 more per person per month because of complete decontrol.) But the study was in many ways based on worst-case assumptions, including the ability of above-market-price contract provisions to withstand market pressures to renegotiate. Others have suggested far smaller price increases that would leave natural gas prices below those of oil. The range of estimates illustrates the difficulties of calculating market prices in advance and, to some extent, the worst fears or fondest hopes of those who produce the calculations.

For the purpose of argument, let us suppose that the higher estimations are correct. The magnitude of the wealth transfer that might occur is star-

---

[5]/"Early Decontrol Could Add $72.8 Billion to Purchase Prices in 1982, INGAA Says," *Energy User's Report,* April 23, 1981, pp. 697–698.

tling, comparable to 10 to 20 percent of the federal budget, an increase of $25 to $45 a barrel of imported oil, or around one-half of all the oil revenues of Saudi Arabia.

However, even if these numbers proved to be accurate, we would still favor a policy of decontrol. The very scale of the numbers illustrates another facet of the issue: The size of the increases that might follow from either partial or complete decontrol is also the measure of how far current prices are from giving the correct information to those making (or not making) investments in natural gas-consuming equipment. Some look at the size of the price increases that might follow decontrol and ask, "How could we possibly pay for natural gas if it were to cost so much?" Others look at the same figures and ask, "How can we ever make this country energy-efficient when natural gas is priced so far below its replacement value?" The investments that are being made every day by consumers, builders, companies, and government itself are based on expectations concerning the price of energy, including natural gas. The net investment every year in material goods that affect energy consumption for years into the future dwarfs the purchase price of energy. Yet, if those investments are made on the basis of incorrect assumptions about the replacement cost of natural gas, many of them will continue to be in place five, ten, or even thirty to fifty years later; and during their entire useful lifetimes, they will have been consuming natural gas at a rate (and at a cost to those who must pay the bill) much higher than is economical. Thus, the cost of deferring a decision on letting the market establish the value of natural gas in the name of protecting the consumer has a long-term effect that is actually harmful to all consumers and damaging to the American economy.

There is another argument for decontrol now rather than later. The world oil market is relatively slack at this time (mid 1982), and inflation rates are drifting downward. Circumstances seem to us as favorable as they are ever likely to be for withstanding the economic shock of decontrol. We cannot be so certain that circumstances will be as good in 1984, 1985, 1987, or any other time in the future.

The magnitudes of the sudden price changes that would follow either partial or total decontrol are large, and users of gas need time to make whatever adjustments those higher prices will require, whether those changes take the form of new investments, doing without, or making partial substitutions. The very best way to make a believable case to everyone that prices are higher is to let them be higher *in fact.* But giving people a definite time when prices will be decontrolled *may* give them time in which to make changes as part of an orderly retreat rather than a rout. Although the experiences under the NGPA give scant reason to be sanguine about the prospect that an announcement of price rises will have the same effect as actual increases, a similar policy was used in that act.

If immediate decontrol is not acceptable, Congress must at least face up to the dangers inherent in decontrol at a leisurely pace. If we cannot cope with immediate decontrol, the law should set a short period of time during which natural gas prices would rise by substantial amounts each year until full decontrol is achieved. Because of the importance of moving prices quickly, it is vital that such a period be truly short.

For example, the schedule for decontrol of new gas might be kept in place with a much higher ceiling price so that the natural gas price would rise in three steps to a 1985 level equaling present estimates of world oil prices, with those estimates being revised twice a year on the basis of current data. At the same time, old gas could be placed on a decontrol schedule (rather than remaining frozen at its 1978 price plus the rate of inflation) that would move it up to a world oil price ceiling followed by total decontrol in 1985.

An alternative method would be to remove price controls on a volumetric basis, so that in each of the next three years, price controls would be removed from one-third of the old gas sold. During that period, the pipelines and the utilities would continue their current practice of rolling in the higher-priced decontrolled gas and the less expensive price-controlled gas. Many believe that in this way the sudden shock of total decontrol would be eased.

Unfortunately, it seems clear to us that almost any method of gradual decontrol that releases part of the gas from control each year will probably result in a sudden price jolt in the first year, rather than spreading out the price rise over time. The price of the decontrolled element of the old gas would probably go high enough so that the average of the decontrolled old gas and the controlled old gas equaled a market-clearing price. What had been planned to be gradual would be sudden instead.

We continue to believe that decontrol of gas prices, whether it is accomplished quickly and all at once or over a believably short period of time, will ultimately be good both for the economy and for individual users, no matter how painful the immediate consequences may be. The failure to act would have effects that, although more subtle and longer range, would be far worse in terms of the health of the economy and the protection of the environment from vast and expensive ventures that can exist only in a protected market.

CHAPTER SIX
COPING WITH OIL EMERGENCIES

Imported oil is in perpetual jeopardy. It is vulnerable to war, revolution, sabotage, and embargo. It is subject to the vagaries of national governments that are allied in OPEC and at the same time allied or split according to religion, culture, language, dynasty, social philosophy, foreign policy, and political ambition. The source of most oil is remote from Western military strength and close to the Soviet Union. More than 15 million barrels a day, half the oil in international trade, exits the Persian Gulf in vulnerable tankers through the narrow Strait of Hormuz.

Disruption, as we have learned, can emerge out of a prolonged crisis or erupt suddenly. Disruption can be piled on top of disruption, war on top of revolution, sabotage on top of war, boycott on top of blockade. When disruption occurs, neither the duration nor the depth of the crisis is predictable. Whatever long-range plan we and other importing nations have for becoming better prepared, there has to be a way of coping with emergencies that come before we are ready.

Gasoline lines and frozen pipes, brownouts and school closings, harvesting emergencies and the closing of energy-dependent businesses are what first come to mind when we think of oil emergencies. But no matter how absorbed we may become in our own domestic crisis, the problem will be more acute for other countries with which our security and our economy are intertwined. We are committed to the security and the independence of Japan and Western Europe and pledged to the survival and economic development of much of the nonindustrial world. Although the United States is the largest oil importer, oil imports are less than half of American oil consumption and less than a fifth of all the energy we consume; in contrast, imported oil provides 80 percent of Japan's total energy. Most industrial countries of Western Europe have little indigenous oil and gas in deposits that can be commercially tapped.

The overriding concern in a severe emergency should, therefore, be disciplined cooperation with the other principal oil-importing countries with which the United States is joined by treaty in the International Energy Agency (IEA). That treaty provides for an orderly sharing of any serious shortfalls in oil supply. It specifies the circumstances in which sharing is to be invoked and the formulas for reduced consumption.

It may come as a surprise to the public that the United States relinquished its freedom to decide unilaterally when, how, and how much to respond to a world oil emergency. But there are compelling reasons for those commitments. One is that a severe oil crisis will be a challenge to national security; any breakdown of the treaty arrangements for meeting an oil emergency will impair the confidence on which the North Atlantic defense community depends. The leadership of the United States is as essential to the one as to the other, and the two are inseparable.

A second reason is that in an emergency, any competitive scramble for oil, any desperate efforts to bypass the sharing system in order to secure preferred access to oil or a share larger than the agreed quota, will simply drive up the price of oil and aggravate the damage. Only discipline on the part of the oil-importing countries can contain an energy crisis and prevent a massive transfer of wealth from consuming to exporting countries.

PRINCIPLES FOR EMERGENCY PLANNING

The microeconomic problem in an emergency will be mainly to accommodate demand to a limited supply. In a crisis, there will be little that can be done, except through inventory management, to augment the supply of oil. An ample strategic petroleum reserve, once accumulated, would be one of the few potential short-term enhancements of supply that may be available. Some substitution among fuels can occur, but the main problem is to reduce demand in an orderly way to fit the estimated supply.

Uncertainty will attend any oil crisis. When a crisis is developing, there will be uncertainty about whether to wait or to take action. There will be uncertainty about whether the crisis will last weeks or years and therefore about the needed durability of any scheme that may be improvised hastily or constructed carefully to cope with it. Uncertainty can generate disputes among IEA members over when to acknowledge the crisis and initiate import controls. There will be uncertainty and dispute about likely responses of supplying countries to proposed actions by consuming countries. And in a crisis worse than any yet experienced, there will be, precisely because experience is lacking, uncertainty about the economic and even the behavioral consequences of an unprecedented domestic emergency.

There are at least five criteria for an optimal response in an oil emergency.

- *The response should contribute to successful international cooperation.* The need for coordinated action can affect both the timing and the extent of measures to curtail imports. Some measures operate directly on the price, with an impact on the quantity of imports that cannot be reliably estimated; other measures operate on the quantity of imports, with an impact on domestic prices that cannot be reliably estimated. The choice may be determined by a commitment to see that imports definitely do not exceed an agreed and specified limit.

- *It should distribute limited supplies of fuel efficiently among domestic uses and users so that the highest-value needs are met.* In the absence of a crisis, misuse of petroleum is merely waste; the highest-value uses can always be met because fuel is available at a price. During a sharp curtailment, it is especially important that the highest-value needs have access to the limited supply. And it is important to avoid the enormous deadweight loss, in time and in fuel, of long waiting lines.

- *It should produce an acceptable distribution of hardships among regions and income levels.* Whatever is done is bound to appear inequitable to some parties, especially if they can think of alternatives that would redistribute the hardships more in their favor. The more an emergency response corresponds to a few principles of fairness, and the less it lends itself to fine tuning or political pressures, the more viable it will be.

- *It should avoid harmful effects on long-term energy policy.* The main danger is in measures that will be politically difficult to remove once the crisis is over.

- *It must accommodate the macroeconomic impact.* There is bound to be an impact—it could be of staggering proportions—on aggregate demand, as well as on the balance of payments, the CPI, short-term capital markets, and exchange rates.

PRICING VERSUS ALLOCATION

It is inescapable that in a severe emergency the federal government will have to restrain the demand for oil. This is a matter of both international obligation and self-interest. Failure to restrain demand would be an invitation to still higher oil prices. The choices are *among* mechanisms for restraining demand and for effecting the internal distribution of petroleum products.

The main options are between mechanisms that work through prices to allow individual choice and mechanisms that are primarily regulatory, prohibitive, and administrative. There is a tendency to believe that prices work well and price fluctuations are moderate when demand and supply change slowly but that direct allocative mechanisms are essential during a sharp curtailment of supply. Sharp increases in the prices of fuels are a visible and measurable hardship; preventing the increases looks like preventing the hardships, and because controlled prices lead to disorderly markets, administrative allocation is needed. The resiliency that prices provide in allocating supplies among competing needs is viewed as a luxury to be abandoned in an emergency.

We believe that point of view is wrong. It underestimates the less visible but more acute hardships that occur when there is no market to meet the myriad urgent needs to which no allocative system can do justice. There is, furthermore, an alternative way to mitigate the hardship and inequity associated with sharp price increases: capture the windfall through taxes and recycle the proceeds in a manner that does not undermine the incentive to reduce oil consumption.

There are two levels at which the price mechanism can be used to restrict demand. One is crude oil, both imports and domestic; the other is refined products (gasoline, heating oil, and other refined products).

The most effective way to limit *imports* of crude oil to a preassigned target, using the price system and capturing the windfall proceeds, would be to sell import licenses to domestic refiners. This should be done by auction, so that refineries have equal access, at the going price, to any desired level of imports. (Releases from the strategic petroleum reserve should be under the same system.) Because all would then pay the same price for imports, there would be no obstacle to passing the costs on to consumers. The higher prices consumers pay for imports would thus be already captured by the government in its auctioning of licenses and could be rebated promptly. The corresponding higher prices of *domestic* oil would be captured by the domestic windfall tax, augmented if necessary, and rebated the same way.

Alternatively, the demand restraints could be applied directly to refined products. Our concern here is not with specifics but with principles. The principles can be illustrated by the rationing of gasoline, a system provided for in legislation toward the end of 1980.

THE MANAGING OF GASOLINE IN A CRISIS

Rationing is not an easy solution to a gasoline shortage. What needs to be emphasized about rationing is how enormous the task is. It is not only the administrative task of distributing rations to more than 100 million recipients (owners of registered vehicles or holders of valid driving permits), compiling the information and getting the mailing addresses correct, avoiding fraud and counterfeit, and handling complaints. It also means determining what the basis of the rationing should be, what persons or occupations or driving activities should be exempt, what formulas are appropriate for adjusting rations to individual needs, and whether rations can be saved for later use. Some passenger cars get twice the mileage other cars get; rural vehicles travel more than urban vehicles; families have up to three automobiles but may have two or three members employed in different locations. People commute short distances and long; some have access to alternative means of transportation, but others do not. There are taxis, delivery trucks, gasoline-burning industrial equipment, emergency vehicles, military vehicles, fishing boats, and farm equipment. Rationing is bound to be crude, arbitrary, awkward, imperfect, and inescapably unfair, and it will be the more so if it is hastily improvised. But the imperfections will be more bearable if rationing is for a brief emergency only; protracted rationing would require procedures for complaints and appeals, enforcement, and politically sensitive comparisons of consumption among states and regions.

A single policy issue is paramount: Should rations be freely marketable? Should buying and selling be not only legal but actually facilitated, with daily quotations printed in newspapers and posted at gas stations?

How would such a market work? Once supplies are curtailed, there will be unsatisfied demand at any controlled price; therefore, coupons will have a price. People who, at the coupon price, want gas in excess of their rations will purchase coupons; people who would rather have extra cash than gasoline will sell some coupons. There should be little difference between the price one customer pays for coupons and the price another obtains for them. There would be interregional trade in coupons and seasonal differences in purchase and sale in different areas. The orderliness of the ration market would depend mainly on the orderliness of the ration system itself—how predictable the rations are, whether rations expire if not used, whether the government changes the gallon value of existing coupons.

The cost of burning gasoline will be the pump price plus the value of the coupon. For the person who buys a coupon at $2 and turns it in to buy a gallon of gasoline for $2, the gallon evidently costs $4. The person who burns six gallons out of a ration of ten, selling four coupons at $2, also spends $4 for each of the six gallons he burns: $2 in cash and a coupon worth $2. Anyone who received ten gallons' worth of coupons while coupons were worth $2 a gallon would soon think of his ration as simply a $20 cash grant, independent of actual consumption.

There are two arguments for marketability. One is obvious: If some people would rather have money in place of some gasoline while others would rather pay and have more gasoline, everybody is better off being allowed to buy and sell coupons. The second is that marketable coupons are far more resilient to the inevitable errors and arbitrariness in the system. With nontransferable coupons, a ration that proves wholly inadequate to the special needs of an individual might be disastrous unless the person could find a way around the law. But with marketable coupons, the damage is limited to some percentage of the market value of the extra coupons needed—a finite amount, and one that is not necessarily huge compared with the urgency of the need.

It is sometimes argued that the well-to-do can have all the gas they want if coupons are marketable and that it is the poor who will convert coupons into cash. But obliging the poor to burn their share of the gasoline when they would rather have the cash simply obliges them to consume expensive gasoline when they would prefer to buy other necessities. It is precisely those who would rather have money than gasoline, because they are poor or because they have comparatively less use for gasoline than for the other things that the money will buy, who are induced by the higher price to conserve gasoline. A nontransferable coupon that might have been worth $2, together with $2 cash, will buy a gallon of gas; a coupon that *is* marketable for $2, together with $2, will buy $4 worth of milk. With marketable coupons, the poor or anybody else can choose to have the gasoline they would have had under the nontransferable system. When they convert coupons to money, they get something they value more than gasoline and simultaneously relinquish gasoline to the market, where there are people who want the gasoline more than they want the other goods and services that $4 would buy.

The system is not much different from a cash benefit coupled with a gasoline tax. A ten-gallon ration with coupons selling at $2 is hardly different from $20 in cash. Gasoline costs $4 a gallon whether a person buys it with his own coupons or with coupons he has purchased. The coupon price is a $2 tax on gasoline, with everybody getting a refundable tax credit equivalent to the "tax" collected.

Thus, except for appearance, the system is equivalent to a $2 tax (or whatever the coupon price is) on every gallon, coupled with a rebate that distributes the proceeds precisely as rations would have been distributed. Whether one received $20 cash or $20 in coupons makes no difference; there is essentially a one-coupon tax per gallon and an unconditional rebate of ten gallons' worth of coupons.

We could save trouble by doing it all with money. Gasoline is distributed through regulated pumps that display the transactions visibly and is already subject to state and federal taxation, so an additional emergency tax would not pose any problem of collection.

Because marketable rations would vary in their value, a genuinely equivalent tax would need to be variable. But the principle is available in what has come to be known as a windfall tax. (Technically, that is an incremental *ad valorem* excise tax, a tax on a fraction of the excess of the sales price over an allowed base price.) At the gas pump, a base price would reflect changes in the cost of crude oil; the tax could then take a high percentage, perhaps 85 percent, of the excess of the pump price over the base price, allowing retention of enough of the price increment to induce competitive behavior at the gasoline pump.

Whatever the mechanics, two features are critical to making a rebate tax equivalent to a marketable ration. One is the synchronized return of proceeds to the consuming public; the second is that the rebates be unconditional, that is, not related to gasoline consumption and therefore not acting as an offset to the higher marginal cost of gasoline. The first is more difficult than the second, but each is in need of careful study as a step toward efficient allocation.

This detailed analysis shows that a pricing mechanism, whether in the form of transferable rations or of a tax-rebate system, can provide the same protection as any administrative formula. But it avoids both the rigidity and the susceptibility to acute error that any allocative system is bound to have. A pricing mechanism minimizes the worst of the hardships associated with nontransferable allocations and is therefore more tolerable politically and administratively. The tax rebate is probably simpler administratively than rationing and capable of being put into effect more promptly.

A GLIMPSE OF A "PRICELESS" ALLOCATIVE SYSTEM

Promptness may be the determining factor. Unless rations had been distributed in advance, it would take months to implement rationing even if preparations were complete. An emergency would probably begin in the Persian Gulf, with a delay of some weeks before tanker arrivals reflected the curtailment. Measures to conserve stocks would be required immediately.

The impact on oil prices would be immediate. Steps to curtail imports would have to be taken immediately. In the absence of controls, domestic prices would rise; and in the absence of a tax or alternative measures such as the auctioning of import licenses, there would be huge transfers of income from consumers to the refineries that received import allocations. The demand for price control would be irresistible. But control without rationing, or without a tax on top of the controlled price, would leave a gap of unsatisfied demand.

Consider just gasoline. There would be federal allocation, mostly by state. There would then be state allocation—or informal allocation by distributors—among local areas and uses; among urban, suburban, and rural areas; and among special or exempt consumers such as taxis, buses, and vehicles operated for emergency, business, farm, state, and municipal purposes. Allocations according to historical use would be crude approximations because annual changes are significant, season and weather affect demand, and prorated reductions within a state lead to such anomalies as supplies of gas in tourist areas but no gas to get there to use it. All these rigidities and awkwardnesses would be piled on top of the inescapable fact that everywhere, in the aggregate and on the average, there would be much less fuel than demand for fuel at the controlled price.

How high might the price have to go to eliminate the unsatisfied demand? There is no reliable estimate because we have no experience with gasoline prices two or three times as high as they are now, and the elasticity of demand calculated from recent experience cannot be extrapolated confidently to prices of $3, $4, or $5 a gallon. But what estimates there are suggest that curtailing consumption by 20 or 25 percent could require a market-clearing price of $3 to $5 a gallon. Adaptation to such prices would undoubtedly moderate demand within months or even weeks and let the price subside somewhat; the higher the price went initially, the greater that effect would be.

Consider now the implications of a market-clearing price of $4 a gallon with the *actual* price held to $2. With gasoline worth *at least* $4 a gallon to anyone trying to buy it and selling at the pump for $2, fifteen gallons delivered into a tank would be worth at least $30 more than what had to be paid in cash at the pump. Someone whose time, at work or at play, is worth $7 an hour (twice the current minimum wage) would be willing to spend at least four hours in line for a tankful. This would be "earning" $30 by exchanging $30 in cash for gasoline worth $60. We have to expect gasoline lines long enough in waiting time for the full cost at the pump (cash plus waiting time) to equal the worth of the gasoline to the consumer.

Calculated on an annual basis (though the situation could not last a year) 60 to 70 billion gallons could entail 5 billion trips to the pump and 20 billion hours of waiting, in stints of four hours or more. At $7 an hour as a modest average value of waiting time, that equals an income loss of $140 billion. This deadweight loss corresponds exactly to the excess of the value of the gasoline over its cost at the pump. The "savings" due to price control get "spent" on access to the gasoline by way of a four-hour wait.

The retailer, whose normal markup is 15 cents a gallon, would be dispensing gasoline on which he could make ten times that much on the black market. Youngsters and others who might ordinarily earn about the minimum wage for performing menial or arduous tasks could hire themselves out at $4 or $5 an hour to do the waiting in line for the rest of us. And if enough of them did it, at hourly rates below what our own time is worth, the lines would simply get longer, until what we paid for having others do our waiting brought the total cost up to that market-clearing price.

At these values, there is no chance that black markets could be avoided. Even if *nontransferable* rations cut the gasoline lines, gasoline itself would be illicitly transferred on a massive scale. As soon as people saw that what they were gaining in the reduced pump price they were losing in waiting time or in hiring people to do the waiting for them, the clamor for an alternative would become irresistible. (The ones who would gain from this situation would be mainly those who made a career out of sitting in the front seats of other people's automobiles.)

This implausible scenario (which omits the bad tempers and the violence) is actually a realistic minimum depiction of what "inefficient distribution" means. The case for some pricing system, even if only in the form of transferable-coupon rations, would quickly become compelling.

We conclude that the market not only is the most feasible and reliable allocative mechanism there is for the stable long run, in which changes occur smoothly, events can be anticipated, and adjustments have time to take place, but also is even more valuable during acute disruption. In a severe oil emergency, crude, awkward, administered allocation can have the most disastrous results precisely because such a situation has no resilience, no flexibility, no safety valves.

THE NEED FOR PREPARATION

We favor pricing over direct allocation in a severe emergency, and among pricing mechanisms, we favor variable taxes coupled with rebates over coupon rationing. We much prefer that the tax be on crude oil, so that all refinery products can be integrated in a single system, without separate schemes for gasoline, heating oil, diesel fuel, industrial feedstocks, and boiler fuel. We do not, however, want to leave the impression that a rebate is simple to devise or easy to implement on short notice or that, compared with marketable rations, everything favors the rebate.

The hypothetical numbers we used to illustrate gasoline lines can be used to suggest the magnitude of likely revenues and rebates. A market-clearing tax on gasoline alone, at those numbers, could entail recycling $150 billion a year—a mammoth fiscal program. Applied to crude oil, to affect all petroleum products, the tax would entail rebating even larger amounts.

That there are alternative ways to recycle proceeds to the tax-paying population or the population of gasoline customers or of fuel consumers is an advantage economically but not necessarily politically. If dispersing the proceeds equally among owners of registered vehicles, as the rationing analogy suggests, were the uniquely appropriate rebate, there would be severe technical problems of assembling data on ownership and some acute impacts on the automobile market to take into account, together with the issues of equitable treatment of families with more than one car and people whose livelihoods require much driving. If the proceeds were rebated in the form of reduced payroll taxes, reduced income tax withholding, or supplementary social security and welfare payments, questions of fairness in the treatment of drivers and nondrivers and multicar families would not arise, but the sheer magnitude of the potential income transfers would guarantee disputes of the most divisive sort. If rebates were channeled to the states in proportion to the estimated proceeds from them, states would differ in their ability to disperse the proceeds promptly through noncontroversial temporary reductions in other taxes.

This need to restore purchasing power by providing prompt rebates is not escaped by leaving *everything* to the market and letting the huge income transfer take place. A price increase of $2 a gallon for liquid fuels would divert consumer purchasing power from other goods to gasoline, heating oil, higher electric bills, and higher transportation costs at an annual rate approaching a quarter of a trillion dollars. Both domestic and foreign suppliers of crude oil will, in the first instance, largely absorb these proceeds or, in the case of a large fraction of the domestic proceeds, pass them on in windfall or corporate income taxes. The macroeconomic shock of such an immense absorption of consumer income would be the greatest ever faced by the U.S. government and would absolutely demand compensatory fiscal and monetary policies devised in advance for the emergency.

Not to prepare the macroeconomic measures (such as a suitable rebate plan) for such a contingency would guarantee an economic disaster in the form of lost income and production far beyond the rise in the direct cost of imported fuel. The immensity of this task, of promptly redistributing tax proceeds amounting to as much as the entire federal income tax, is exceeded only by the enormity of the income transfers that could occur without any rebate system at all or of the waste and disorder that would accompany price controls alone.

Clearly, there is a need for preparation. Even if the political decisions cannot be made in advance, the choices can be identified, the impacts of alternative schemes can be assessed, and the legal and technical problems can be anticipated. Perhaps only the onset of a genuine emergency, not a hypothetical emergency selected for planning purposes, could motivate resolution of the inevitable controversies. The ultimate choices may have to be determined by what can be erected in a hurry.

We reach three conclusions. First, in a severe emergency, the income transfers both domestically and to foreign suppliers could be of such forbidding magnitudes that there would have to be intervention by the federal government to mitigate the macroeconomic shock and to prevent a huge redistribution of income. Second, price controls, direct allocations, and the suspension of the price system where gasoline and other petroleum products are concerned would entail intolerable waste, hardship, and disorderliness. Third, when a severe emergency strikes, there may be little time to take the preparatory steps essential to implementation of any satisfactory market pricing system, whether it is in the form of rebates coupled with taxes or of marketable rations. Therefore, in the very interest of market pricing in such an emergency, we urge that planning go forward.

MEMORANDA OF COMMENT, RESERVATION, OR DISSENT

By JACK F. BENNETT, with which RODERICK M. HILLS has asked to be associated, regarding Chapter Five

The comments on pages xvi and 65 are much too weak in their condemnation of a decontrol tax and would make matters worse by implying that such a tax should be permanent on all old gas until it stops flowing. A decontrol tax, even if restricted to old gas, would hinder the application of novel techniques to maximize recovery of additional gas from old gas fields and would reduce overall productivity through regulatory complexity. If such a tax were imposed, it should be phased out as soon as possible rather than persist until the gas was depleted.

By DON C. FRISBEE, with which J. W. McSWINEY and JOHN F. WELCH have asked to be associated, regarding Chapter Four

This chapter is appropriately concerned with the inadequacy and, in most cases, the misleading result of traditional electric-utility pricing as a means of conveying to consumers the resource cost to society of increased electrical consumption. This defect has grown in size and significance in recent years due to the abrupt shift in the electric-utility industry's basic economic position, from one of declining unit cost to rapidly increasing unit cost. As a consequence the industry's marginal costs now generally exceed marginal revenues.

While economic theory would seek a pricing mechanism which clearly identifies the cost of new electric supplies to consumers, the industry (and for similar as well as different reasons, its regulators) have other problems created by the circumstances which have produced this serious pricing defect. Generally speaking, an abrupt shift to replacement-cost pricing would enlarge many of those problems with disruptive and dangerous results.

Wholly apart from the desirability of pricing electricity more in line with its true economic costs, there is an underlying need to strengthen the earnings and borrowing capacity of utilities. There are specific practical examples of various electricity pricing changes that could, under current circumstances, work in the same direction as these general principles. These changes could and should be considered and acted upon relatively quickly in view of the financial situation facing much of the electric-utility industry.

An obvious first step in this direction would be the inclusion of construction-work-in-progress (CWIP) in the utility rate base. Such a move would provide a partial change in rates in the direction of replacement-cost pricing. A welcome complement of the change would be the sharp reduction in future cost-of-replacement capacity due to the collection of financing charges currently, as construction proceeds, rather than their capitalization and addition to the cost of future plant.

Many state regulatory commissions already include a portion of CWIP in rate base. The Federal Energy Regulatory Commission permits inclusion of CWIP in rate base under certain conditions and is currently holding hearings on the matter. The advantage of moving in this direction is that a procedure for including a small portion at first, and an ever-growing proportion as time goes by, of CWIP in the rate base will provide an opportunity for gradually feathering in the full amount—a procedure which may well be more acceptable to the public at large.

Such a step would move electricity prices significantly in the direction of replacement costs and would do so without creating additional net earnings for the stockholder—except indirectly as such earnings might result from a larger cash flow and a healthier financial condition of the firm.

Besides the above major change, there are several smaller-scale changes in rate-setting policies and procedures that could work in the same direction if adopted by the relevant regulatory agencies. They include: (1) the adoption of forward-looking test years in regulatory proceedings; (2) the establishment of rates of return based on current money costs, rather than using only "imbedded" interest rates; (3) prompter action on rate applications; and (4) more frequent granting of interim relief when the full rate proceedings are expected to be protracted. These would be practical improvements in rate-setting actions that could help to make their results more timely. None of the above changes would be the same thing as replacement-cost pricing, but under current circumstances they would work constructively in that same direction.

By DON C. FRISBEE, with which J. W. McSWINEY has asked to be associated, regarding Chapter Four

Another step that is badly needed is the adoption of significantly higher depreciation charges on utility plants—charges that could more nearly offset the effect of inflation on the cost of plant replacement. Indeed, such more-realistic depreciation charges could be high enough to offset the increased gross receipts from higher energy prices described here; in that case, the so-called revenue-overcollection problem would be nonexistent. One approach to higher depreciation would simply be to require that all tax depreciation provisions be booked and recognized for rate-making purposes. Another approach that deserves serious study—and one more consistent with the philosophy of replacement-cost rate-making—would be to permit depreciation provisions for rate-making purposes to be made on a last-in first-out basis. Thus, each year the newest and probably the most expensive plant would be depreciated first. Such a procedure—as a manner of coping in part with the effects of inflation, for both financial reporting and rate-making purposes—would be consistent with the LIFO system of inventory accounting which has been in effect for years and has served a similar purpose in the inventory accounting field; i.e., it has helped to keep paper profits on inventory write-ups from distorting operating earnings.

By GEORGE C. McGHEE, regarding Chapter Five

I believe that phased deregulation of natural gas prices, which succeeded in oil and has been accepted in principle by the Cabinet Council on Natural Resources, is preferable to the drastic across-the-board deregulation recommended in this report. All gas discovered in the future should be deregulated, as recommended by the Council. Congress will strongly oppose changes in the present law that deregulates all but "old" gas by 1985, particularly without a "windfall-profits" tax, which is widely accepted by industry but opposed in this report. Benefits cited in the report do not appear to justify such drastic increases in consumer prices (estimates go up to $72.8 billion the first year) while the country is still fighting inflation, with the resulting hardship to lower income groups. Immediate deregulation would result in no appreciable increase in gas supplies (the report cites a possible decrease), which in any event are adequate for several decades. Gas reserves, which stand at 10 years consumption, declined only one percent in 1980. Increase in gas prices, which consumers already know is inevitable by 1985 and must be prepared for, will discourage conversion to gas from scarce oil and delay savings in oil imports.

By J. W. McSWINEY, with which JOHN F. WELCH has asked to be associated, regarding Chapter Four

I am concerned that replacement-cost pricing of electricity is too theoretical and long-term a concept to be of much help in the current situation facing the electrical-energy industry. I believe it will be difficult for public-utility commissions and electricity users to accept this concept and the pricing consequences that would flow from it. It would be unfortunate if debate over this idea were to divert attention from the urgent need for public utilities to be allowed a rate of return on their invested capital that is competitive with that earned in other industries seeking capital in the marketplace.

In my view, failure to achieve that level of return is in many cases a major handicap to utilities seeking to raise sufficient funds to honor their mandate to provide needed electricity over the years ahead.

By ROGER B. SMITH, regarding Chapters Two, Four, and Five

While supporting many of the free market concepts appearing in the executive summary, I disapprove of many of the positions dealing with Energy Pricing Policy and Low-Income Consumers, Electricity Pricing Policy, and Natural Gas Pricing Policy.

By WILLIS A. STRAUSS, regarding Chapter Five

While I believe strongly in a "free market" economy, I cannot accept the basic conclusion of the paper that total and immediate decontrol of natural gas is appropriate after 25 years during which federal policy held the price of natural gas below its true market value. The shock to an already fragile economy would likely be too great. The immediate decontrol of all natural gas wellhead prices would severely harm the American economy.

I am not convinced that decontrolling old flowing-gas prices will prompt additional exploration and production. A more likely outcome to such a move would be the transfer of wealth from consumers to producers and royalty owners. There is no assurance that such a wealth transfer would result in investment in exploration efforts. I do not believe we should burden the consumer with such an income transfer on the vain hope that the Natural Gas Policy Act has provided adequate economic incentives without decontrol of old flowing gas.

If old flowing-gas wellhead prices are not decontrolled, there is no need for an excise or "windfall profits" tax. I strongly oppose the imposition of any such tax since I feel it would provide a disincentive to exploration and would unduly burden the consumer. Further, no assurance could be given that the proceeds of an excise tax could be channeled back to the consumer in any equitable fashion.

The paper states that pipelines would suffer no immediate adverse economic effects from higher prices since they are permitted full pass-through. I cannot accept that conclusion. This is evidenced by the questions being raised before the Federal Energy Regulatory Commission as to the prudency of pipelines paying high prices for currently decontrolled natural gas.

I am also concerned about the effects indefinite escalation clauses may have on wellhead prices upon decontrol. If, as I believe, the very high prices being paid for currently deregulated gas will trigger those clauses, the effect will be a severe price fly-up which will have a severely disruptive effect on the economy and which may increase imports of oil with further serious economic effects.

Our movement toward market-oriented pricing, while an extremely desirable goal, must come without the severe market disruption which would be attendant to immediate and total decontrol. Deliberate, phased movement toward the goal of market-oriented pricing of newer gas, retention of controls on old flowing gas, and a continuation of rolled-in pricing is an appropriate means of moving toward the interrelated goals of reducing our dependence on imported energy, developing our domestic energy sources and increasing economic stability.

OBJECTIVES OF THE COMMITTEE FOR ECONOMIC DEVELOPMENT

For forty years, the Committee for Economic Development has been a respected influence on the formation of business and public policy. CED is devoted to these two objectives:

To develop, through objective research and informed discussion, findings and recommendations for private and public policy which will contribute to preserving and strengthening our free society, achieving steady economic growth at high employment and reasonably stable prices, increasing productivity and living standards, providing greater and more equal opportunity for every citizen, and improving the quality of life for all.

To bring about increasing understanding by present and future leaders in business, government, and education and among concerned citizens of the importance of these objectives and the ways in which they can be achieved.

CED's work is supported strictly by private voluntary contributions from business and industry, foundations, and individuals. It is independent, nonprofit, nonpartisan, and nonpolitical.

The two hundred trustees, who generally are presidents or board chairmen of corporations and presidents of universities, are chosen for their individual capacities rather than as representatives of any particular interests. By working with scholars, they unite business judgment and experience with scholarship in analyzing the issues and developing recommendations to resolve the economic problems that constantly arise in a dynamic and democratic society.

Through this business-academic partnership, CED endeavors to develop policy statements and other research materials that commend themselves as guides to public and business policy; for use as texts in college economics and political science courses and in management training courses; for consideration and discussion by newspaper and magazine editors, columnists, and commentators; and for distribution abroad to promote better understanding of the American economic system.

CED believes that by enabling businessmen to demonstrate constructively their concern for the general welfare, it is helping business to earn and maintain the national and community respect essential to the successful functioning of the free enterprise capitalist system.

OBJECTIVES OF THE CONSERVATION FOUNDATION

The Conservation Foundation is a nonprofit research and communications organization based in Washington, D.C. The Foundation's primary purposes are to improve the quality of the environment and to promote wise use of the earth's resources. The Foundation achieves these purposes by conducting interdisciplinary research and by communicating its views and findings to policy makers and opinion leaders. The Conservation Foundation does not have members, is not a lobbying organization, and does not buy or sell land.

Since its founding in 1948, The Conservation Foundation has been an advocate characterized by reason and balance, recognizing the importance of a healthy social and economic climate to the achievement of conservation goals. The Foundation believes that public policies must be based on rigorous factual analysis and public understanding.